$ 30.00

Tropical Tree Fruits for Australia

ussie tree nay go back 0,000 years

ŝOCIATED PRESS

SYDNEY, Australia — A sprawl-
pine tree on a rainy mountain in smania could be more than ,000 years old, making it one of e world's oldest living things, a wspaper reported today.

"It is certainly plausible that we e looking at a 10,000-year-old ganism," said Ed Cook, a Columbia University tree-dating specialist oted in *The Sydney Morning Herald*.

"It represents the entire record-history of humanity."

The tree was found by forestry ficial Alan Peterson in 1985 near e summit of 3,795-foot-high high ount Read on Tasmania, the island off Australia's southern coast. ount Read, in the northwest of e island, receives 13 feet of rain nually.

The tree is a mass of tangled ood and roots covered in thick oss and spread over a 2.5-acre ea.

Peterson initially thought it was small forest of individual trees. ut testing showed that every leaf, ranch and root had the same DNA tructure, indicating they were part f the same organism.

Ancient pollen from the tree, ound in samples taken from a nearpy lake bed, is 10,500 years old, *The Herald* reported.

The oldest known living organism is a creosote bush in the Mojave Desert thought to be 11,700 years old.

Queensland Department of Primary Industries
Information Series QI83018

Tropical Tree Fruits for Australia

Compiled by
P. E. Page
Horticulture Branch

Queensland Department of Primary Industries
Brisbane 1984

National Library of Australia Cataloguing-in-Publication Data

Tropical Tree Fruits for Australia

Includes index

ISBN 0-7242-2200-6

I. Tropical tree fruit — Australia

I. Queensland Dept. of Primary Industries

II. Title. (Series: Queensland Department of Primary Industries Information Series: QI83018).

635 0489238749

ISSN 0727-6273

First Published May 1984
Reprinted October 1984 with alterations
Reprinted July 1985

Queensland Department of Primary Industries

GPO Box 46

Brisbane 4001

Typeset and Printed by Poly-Graphics Pty. Ltd., Brisbane.

Some of the mouth-watering fruits discussed in these papers. Photograph courtesy of Mr. J. Truscott, Cairns. Fruit kindly supplied by B.J. Watson, Kamerunga Horticultural Research Station and fruit-growing enthusiasts in north Queensland.

FRUITS OF NORTH QUEENSLAND

	Common name	Botanical name
1	Pineapple	Ananas comosus
2	Durian	Durio zibethinus
3	Jackfruit	Artocarpus heterophyllus
4	Honeydew Melon	Cucumis melo
5	Rambutan	Nephelium lappaceum
6	Lychee	Litchi chinensis
7	Red Pitaya	Hylocereus undatus
8	Banana	Musa spp.
9	Watermelon	Citrullus lanatus
10	Breadfruit	Artocarpus altilis
11	Mango	Mangifera indica
12	Avocado	Persea americana
13	Yellow Passionfruit	Passiflora edulis f. flavicarpa
14	Guava	Psidium guajava
15	Pummelo	Citrus maxima
16	Persimmon	Diospyros kaki
17	Guanabana (Soursop)	Annona muricata
18	Papaya (Papaw)	Carica papaya
19	Kitembilla	Dovyalis hebecarpa
20	Abiu	Pouteria caimito
21	Mangosteen	Garcinia mangostana
22	West Indian Lime	Citrus aurantifolia
23	Black Passionfruit	Passiflora edulis
24	Bell Fruit	Syzygium aqueum
25	Macadamia	Macadamia tetraphylla
26	Chiku (Sapodilla)	Manilkara zapota
27	Carambola	Averrhoa carambola
28	Jelly Palm	Butia capitata
29	Sweetsop (Sugar Apple)	Annona squamosa
30	Longan	Euphoria longan
31	Prickly Pear (Tuna)	Opuntia ficus-indica
32	Monstera	Monstera deliciosa
33	Rockmelon	Cucumis melo
34	Atemoya	Annona atemoya Hort.
35	Cashew	Anacardium occidentale
36	Acerola	Malpighia glabra
37	Uvilla	Pourouma cecropiaefolia
38	Tamarillo	Cyphomandra betacea
39	Casimiroa (White Sapote)	Casimiroa edulis
40	Mabolo	Diospyros discolor
41	Cherimoya	Annona cherimola
42	Black Persimmon	Diospyros digyna
43	Ambarella (Kedondong)	Spondias cytherea
44	Icaco (Cocoplum)	Chrysobalanus icaco

FOREWORD

In September 1980 I was privileged to officially open the first Australian Tropical Tree Fruits Workshop and in so doing saw an opportunity for the publication of a book of this type.

I suggested that my Department might offer to undertake this enterprise and the workshop participants agreed that the proceedings would form the basis of the book and that it should be illustrated in colour.

The project has now been completed. This book should provide a useful information source on the production of a range of fruit, many of which are new to the Australian scene. Their exotic new tastes and appearances will widen the sensory experience of the Australian public, and open their eyes to the incredible diversity of fruits that are available to our neighbours in the tropics.

Research on the production and post-harvest handling of these fruits is expanding within my Department and other kindred organizations, in order to meet the increasing public demand for and interest in these fruits.

I am sure this book will prove to be a useful reference to all interested in the production and use of tropical fruit.

Dr G. I. Alexander
Director-General,
Queensland Department of
Primary Industries.

CONTENTS

LIST OF PLATES

Colour Section 1 (between pages 48 and 49)

Colour Section 2 (between pages 80 and 81)

Colour Section 3 (between pages 160 and 161)

Colour Section 4 (between pages 208 and 209)

AUTHORIAL CONTRIBUTORS

D. McE. Alexander, CSIRO, Division of Horticultural Research, Merbein, Vic.

D.J. Batten, New South Wales Department of Agriculture, Tropical Fruit Research Station, Alstonville.

P.R. Beal, QDPI, Redlands Horticultural Research Station, Ormiston.

K.R. Chapman, QDPI, Maroochy Horticultural Research Station, Nambour.

B.W. Cull, QDPI, Maroochy Horticultural Research Station, Nambour.

P.J. Farlow, QDPI, Redlands Horticultural Research Station, Ormiston.

M.A. Foale, CSIRO, Division of Tropical Crops and Pastures, Cooper Laboratory, Lawes, Queensland.

A.P. George, QDPI, Maroochy Horticultural Research Station, Nambour.

K.R. Jorgensen, QDPI, Maryborough.

D.H. Maggs, CSIRO, Division of Horticultural Research, Merbein, Vic.

I.L. Miller, Northern Territory Department of Primary Production, Darwin.

J.V. Possingham, CSIRO, Division of Horticultural Research, Adelaide.

P.B. Scholefield, CSIRO, Division of Horticultural Research, Darwin.

M. Sedgley, CSIRO, Division of Horticultural Research, Adelaide.

J.D. Sturtz, Northern Territory Department of Primary Production, Darwin.

B.J. Watson, QDPI, Kamerunga Horticultural Research Station, Cairns.

A.W. Whiley, QDPI, Maroochy Horticultural Research Station, Nambour.

SLIDE CONTRIBUTORS

The following individuals and organizations contributed the colour slides from which the colour plates were made.

D.J. Batten, Tropical Fruit Research Station, Alstonville, N.S.W. — plates 1, 35, 39, 52, 58.

A. Carle, Cairns — plates 5, 9, 13, 15, 18, 19, 24, 25, 26, 27, 31, 36, 37, 43, 45, 50, 64, 65, 66.

K.R. Chapman, Maroochy Horticultural Research Station, Nambour — plate 54.

CSIRO Division of Horticultural Research — plates 2, 14, 16, 30, 40, 49, 51, 62, 68.

B.W. Cull, Maroochy Horticultural Research Station, Nambour — plates 7, 8, 11, 17, 32, 38, 46, 47, 48, 53, 57, 67.

P.J. Farlow, Redlands Horticultural Research Station, Ormiston — plates 42. 44.

A.P. George, Maroochy Horticultural Research Station, Nambour — plates 4, 6.

P.E. Page, QDPI, Brisbane — plates 20, 21, 22, 23, 41, 60, 61, 63.

B.J. Watson, Kamerunga Horticultural Research Station, Cairns — plates 3, 10, 12, 28, 29, 33 34, 55, 56, 59.

PREFACE

This book contains the crop review papers prepared for the Australian Tropical Tree Fruits Workshop held at Marcoola Beach, Queensland in September 1980. Additional reviews are included on macadamia, avocado and pummelo. These crops were not discussed at the Workshop.

The Workshop, the first of its kind held in Australia, was approved under the A.A.C. Technical Conference Programme, and was jointly organized by the CSIRO Division of Horticultural Research, the Queensland Department of Primary Industries, and the New South Wales Department of Agriculture.

The 5-day Workshop reviewed the horticultural characteristics of approximately 50 tropical fruit and nut crops, and assessed their likely potential for further development in Australia. Present major tropical fruit crops such as avocados, macadamias, bananas, papaw and citrus were not discussed.

The review papers were prepared by officers of the CSIRO Division of Horticultural Research, the Queensland Department of Primary Industries, the New South Wales Department of Agriculture, the Northern Territory Department of Primary Production, and the CSIRO Division of Tropical Pastures. The Queensland Department of Primary Industries has published this book on behalf of the Workshop and gratefully acknowledges the contributions made by the many authors involved, and the permissions given by their organizations to allow this material to be presented here.

The colour photographs used to illustrate the text were collected from numerous sources throughout Australia. The Queensland Department of Primary Industries gratefully acknowledges the efforts made by many people in contributing colour photographs for use in the book.

The resolution to publish these review papers in book form was passed at the Australian Tropical Tree Fruits Workshop. It is believed that the information contained herein has not previously been assembled under one cover, and that the book is unique in detailing the characteristics of a large number of tropical tree fruit and nut crops with potential for future development in Australia.

Chapter 1

POTENTIAL FOR FRUIT GROWING IN TROPICAL AUSTRALIA

by D. McE. ALEXANDER and J.V. POSSINGHAM

Historically, fruit growing in Australia and tropical fruit growing in particular, has spread with development and has initially been concentrated near centres of population. Most production of tropical fruits has come from the coastal plains of the east coast extending from about Coffs Harbour in the south to about Maryborough in the north.

Most fruit produced further north are consumed locally. The exceptions are bananas from around Tully, mangoes from around Bowen, papaws from the Yarwun district, and more recently avocados from the Atherton Tableland.

The principal fruit crops grown in Australia are grapes, apples, citrus, peaches, bananas and pears. Tropical fruit crops which are grown to only a limited extent include pineapples, papaws, avocados, mangoes, custard apples and passionfruit with even smaller quantities of persimmons, mulberries, pomegranates, guavas, feijoas, tree tomatoes, Chinese gooseberries (kiwifruit) and dates. While all the major crops are to some extent dependent on export markets tropical fruits are consumed entirely within Australia. Exceptions are bananas and preserved pineapples. In addition to commercial crops there are significant plantings of household or backyard crops. It has been estimated that these average 2.5 trees per household.

Within the tropics, which extend from latitude 23°27' north to 23°27' south of the equator, there are many different climates. Proximity to the sea, sea currents, altitude, direction of the prevailing winds, rainfall and air humidity all have modifying effects.

Other things being equal air temperature is modified by elevation where a lapse rate of about 5.5°C per 1000 m above sea level might be expected and by distance from the equator where a lapse rate of about 1°C per 1000 km south is appropriate. High humidity has a modifying influence particularly in the wet season.

The mean monthly maximum and mean monthly minimum temperatures for Australia in January (the hottest month) and July (the coldest month) are shown in figures 1,2,3 and 4 respectively. (see end of chapter)

The cultivation of strictly tropical species is limited to those areas where frost never occurs (figure 5). However, a probability of frost occurring in one year in twenty (say) might still be acceptable whereas an occurrence in one year in five is not. Hence potential areas could extend closer to the 300 frost-free day line if other factors were suitable. Of the total area of Australia (about 7.7×10^6 sq km), 3.1×10^6 sq km, or about 40%, is in the tropics but only about half of this occurs in the frost-free zone. This includes a narrow coastal strip south of the tropics on both east and west coasts.

A major consideration in determining the suitability of any site for horticulture is a water supply, and here rainfall with its attendant air humidity is of utmost importance (figure 6). Apart from small areas in the vicinity of Cairns which have only a very short dry period, tropical Australia experiences a dry of from 5 to 7 months' duration. Evaporation from a standard pan varies from around 150 cm near the east coast to over 360 cm per year in the driest regions. In figure 6 the 50 percentile median rainfall is presented. This is probably more relevant than the average rainfall because adjustments are made for unusually high and unusually low rainfall years. Figure 7 shows rainfall zones based on median annual rainfall and the seasonability of the rain.

Apart from crops such as mangoes, some annonas, dates, pistachios and ziziphus in some isolated areas, irrigation is an essential requirement for horticulture in the Australian tropics.

Suitable soils for horticulture occur on levees of the major rivers throughout the region. Most recommendations for fruit growing require at least 1 m deep, well-drained soil but topography and special microclimate conditions may make shallower soils equally productive. Layers in the soil with a high clay content or of laterite can impede drainage to such an extent that many species would not survive. However, some species such as *Ziziphus, Garcinia* and *Phoenix* can withstand some waterlogging.

Wind is often a major hazard for horticulture and most of our tropical coast is subject to cyclones. The frequencies of occurrence of peak 3-second wind gusts 10 m above an open field have been calculated for Roeburne Shire, Hedland, Darwin, Cairns, Townsville, Mackay (Anon. 1980) and show that peak 3-second gust speeds of 200 km per hour are likely to occur about once in 100 to 200 years while speeds of 100 km per hour are likely to be exceeded once in about 10 years at each of these centres. While wind speeds diminish with distance from the coast effects may be severe up to 50 km inland and further inland associated flood rains may also be damaging to horticulture. The use of wind-breaks and protected situations are the horticulturist's best means of defence.

Markets

The 1975 National Population Inquiry concluded that the human population of Australia will increase by no more than 1.5% per annum so that by the year 2000 the population will have reached no more than 20 million people. Population shifts similar to those which have occurred with mining activities for iron ore, coal and uranium are likely to continue.

Potential markets for tropical fruits within Australia must therefore be in competition with traditional subtropical fruits unless export markets can be developed. Australian exports to and imports of fruits from countries with over $10,000,000 of trade for 1977–78 are shown in figures 8 and 9. These illustrate the relatively small value of tropical fruit exports and the dominance of nut crops over other fruit imports.

Tropical fruit growing overseas

No tropical fruits are grown in Europe and relatively small amounts in the United States, the two main sources of Australian knowledge about fruit growing. The United States grows tropical fruits in Florida where the availability of suitable soils and occasional below-freezing winds limit production. Hawaii grows a range of tropical fruits but has a most unusual maritime tropical climate that is modified by altitude. For this reason varieties of both papaya and macadamia developed in Hawaii have complex environmental requirements and have had limited success when grown in other tropical countries. In recent years the United States has invested in tropical fruit growing in a number of countries, growing bananas in Honduras, Puerto Rico and Ecuador; pineapples in Taiwan and Kenya; West Indian limes in Mexico; and oranges in Brazil. This has led to an increase in the flow of traditional and scientific knowledge about tropical fruit production between these countries, the United States, and the rest of the English-speaking world.

Small quantities of tropical fruits for European markets are grown in Israel where desert, and often saline, areas with low humidity and elevation are irrigated to generate unusual growing conditions for these crops. Crops include bananas, pineapples and avocados.

Many equatorial tropical countries, including those on the continents of Africa, South America, the Indian subcontinent, and South East Asia are third world or less developed countries. In these regions improvements to traditional methods for the production of tropical fruits were earlier on made by expatriate Europeans, and more recently by FAO/UNDP specialists. However, there is no major long-standing international institute in the tropics for research on tropical fruits. Urgent research is needed on variety improvement, methods of propagation, and the control of fungal, viral and insect diseases, nutrition, post-harvest handling and storage. A specific problem in countries located close to the equator is the uniform climate throughout the year, which can lead to continuous vegetative growth, a series of small crops each year and low quality fruit.

Grapevines are an emerging crop of importance in many tropical countries where relatively high yields of table grapes can be obtained. These commonly bring high prices on local markets. Double cropping is common and pruning techniques aim for cropping at times of the year when there is little or no rain. Fungal diseases on the leaves and fruit are major problems and limit the production of pure *Vitis vinifera* varieties. The long-term future of grape growing in the tropics is dependent on the breeding of mildew-resistant hybrid varieties using a range of *Vitis* species.

There is a high demand for tropical fruit in Europe with supplies coming from former European colonies in Africa and South America and from Israel and South Africa. In South Africa considerable effort has been put into the development of a tropical fruit industry for export. Many third world countries are unable to export significant quantities of fruit because the high local demand due to population increases, keeps abreast of any increases in production.

Much of the tropical fruit of the world is still produced in the greatest quantity and of the best quality at or near to the original centres of origin of the species. For example, the range and quality of mangoes is high in India, lychees excel in southern China in the area close to Canton, and mangosteens, durians and salak are largely confined to the Malaya-Indonesia region. By contrast, pineapples commonly those of the cayenne type, bananas mainly of the dwarf cavendish variety, and papayas of the solo type, are almost ubiquitous in the tropics. High altitude areas near to the equator in countries such as Kenya, Colombia, Peru, Nepal and India grow a surprisingly wide range of tropical fruits including bananas, papayas, pineapple, macadamia, at altitudes of 2000 to 2500 m. In nearly all cases these fruits have found unusual environmental niches which are frost free and have an aspect such that they receive a high level of solar radiation. In these situations cultivars which have no specific cold tolerance are commonly used.

Tropical fruit production in Australia could be developed to cater for either the domestic or the export markets. The former is limited in size but could cater for small quantities of a large number of species. If we are to successfully exploit the export market, however, we will probably need to concentrate on a selected few species, preferably those with a longer post-harvest life than average or that are suited to processing. As well, although our southern hemisphere location will confer some advantages, there may be only a limited opportunity for export sales unless Australia is able to intensify production and mechanize more of the harvesting and handling operations. Increased research on crop production, handling and marketing is necessary.

References

Anon. (1980), Cyclones — assessing the risk. *Ecos* **23,** 7.
Australian Bureau of Meteorology (1975a), Climatic Atlas of Australia — Temperature. AGPS, Canberra.
Australian Bureau of Meteorology (1975b), Climatic Atlas of Australia — Rainfall. AGPS, Canberra.

Further reading

Tribe, D.E. (1980), Animal production in the 21st century. *Journal of Australian Institue of Agricultural Science* **46,** 36.

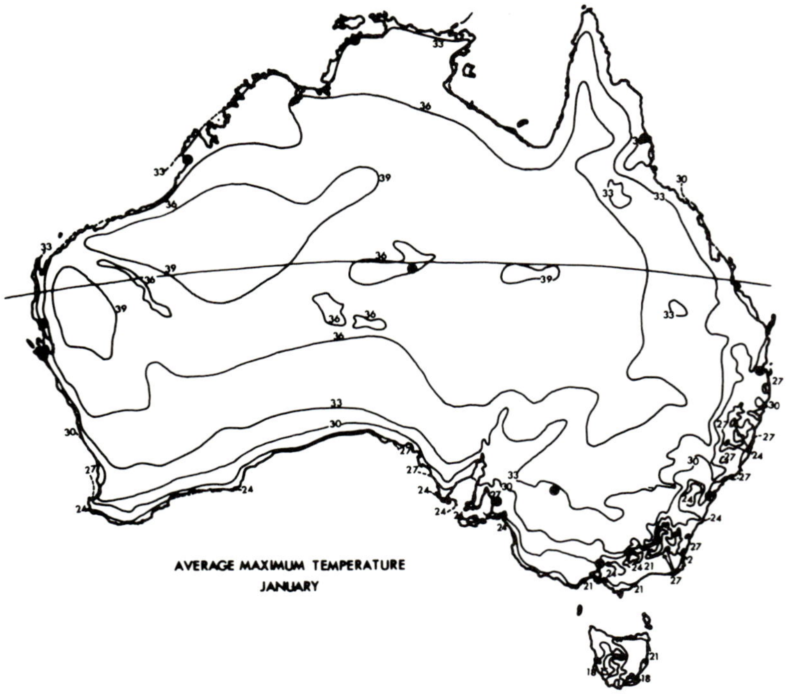

Figure 1: Average maximum temperature for January, the hottest month.

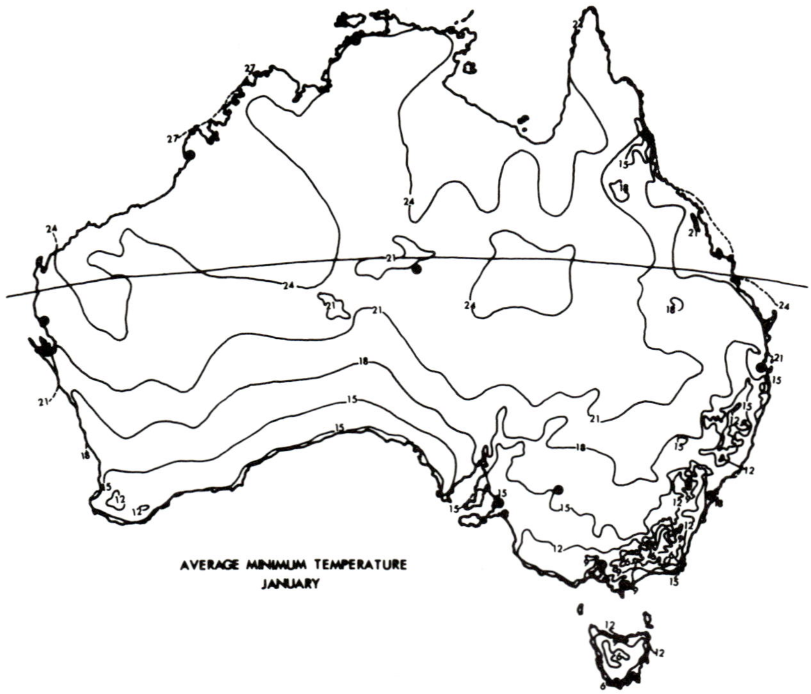

Figure 2: Average minimum temperature for January.

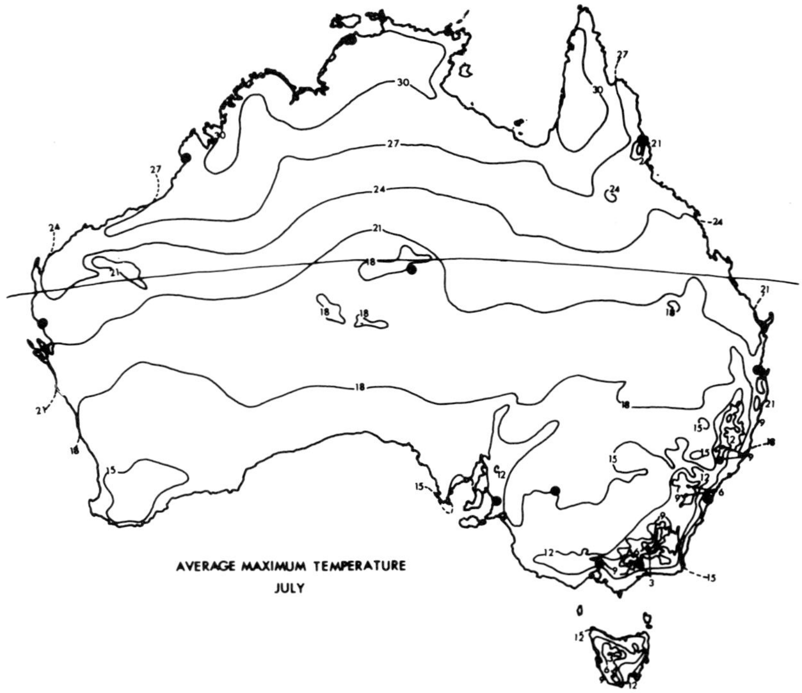

Figure 3: Average maximum temperature for July, the coldest month.

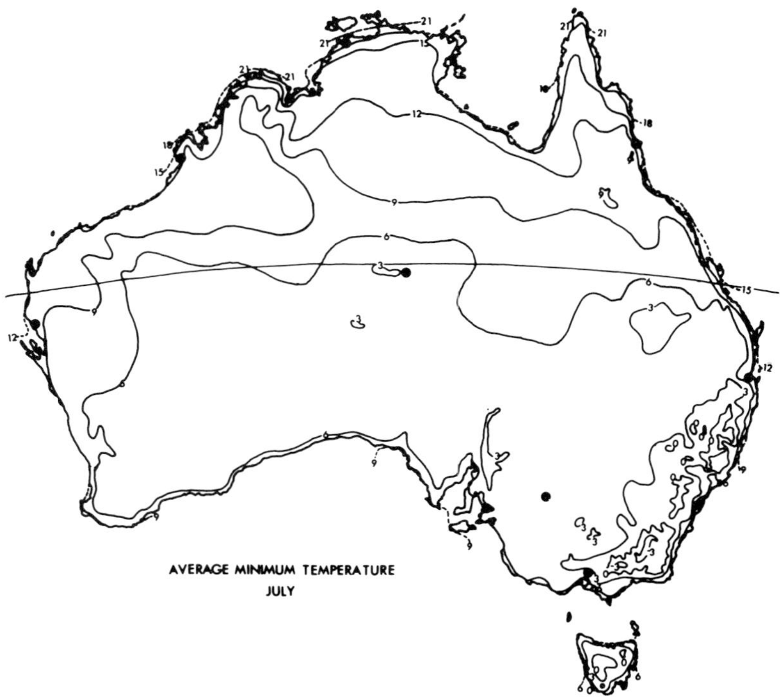

Figure 4: Average minimum temperature for July.

Figure 5: Average frost free period.

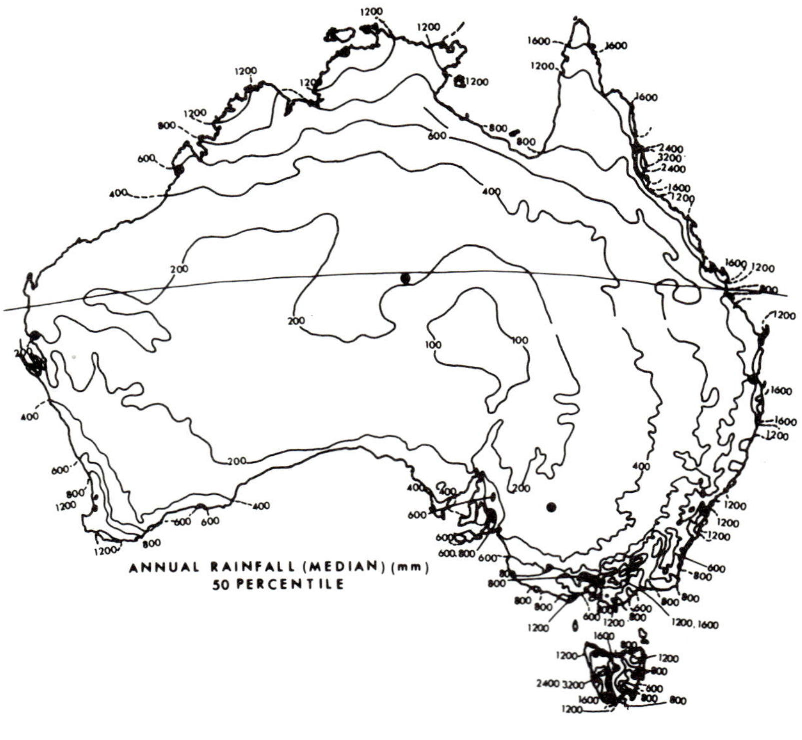

Figure 6: Annual rainfall (mm) (Median); 50 percentile.

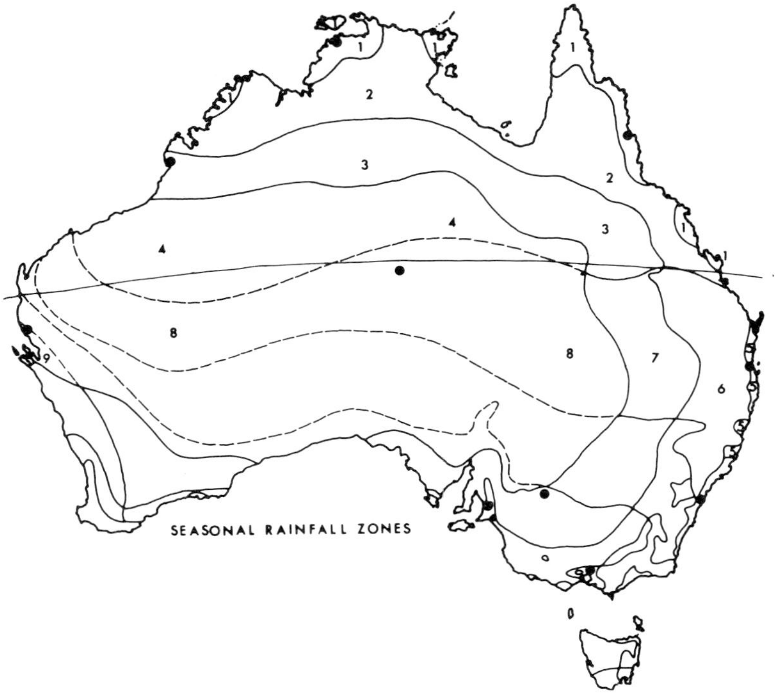

Figure 7: Seasonal rainfall zones.
Legend

Zone	*Rainfall (mm)*	*Season*	*Seasonal Ratio*
1	> 1200	Summer dominant	>3.0
2	650-1200	Summer	>1.3 - 3.0<
3.	350-600	Summer	>1.3 - 3.0≤
4.	< 350	Arid zone Summer dominant	>3.0
5.	< 1200	Summer	>1.3 - 3.0≤
6.	650-1200	Summer	>1.3 - 3.0≤
7.	350-600	Summer	>1.3 - 3.0≤
8.	< 350	Arid zone	>1.3 - 3.0 ≤
9.	< 250	Arid zone Winter dominant	>3.0

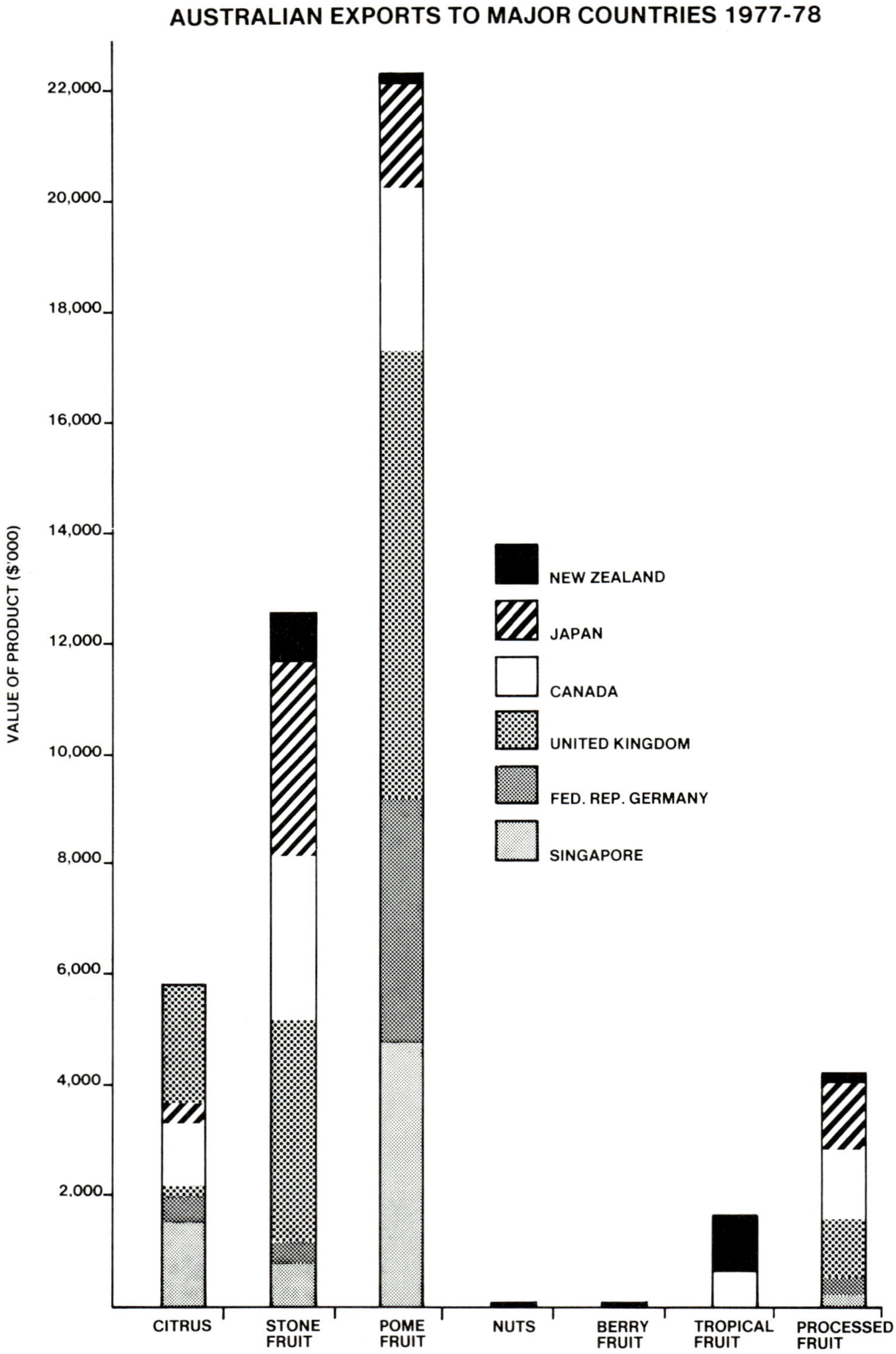

Figure 8: Australian exports for 1977-78 of fruits (excluding vine fruits) to countries for which exports of fruits and vegetables were valued at more than $10,000,000.

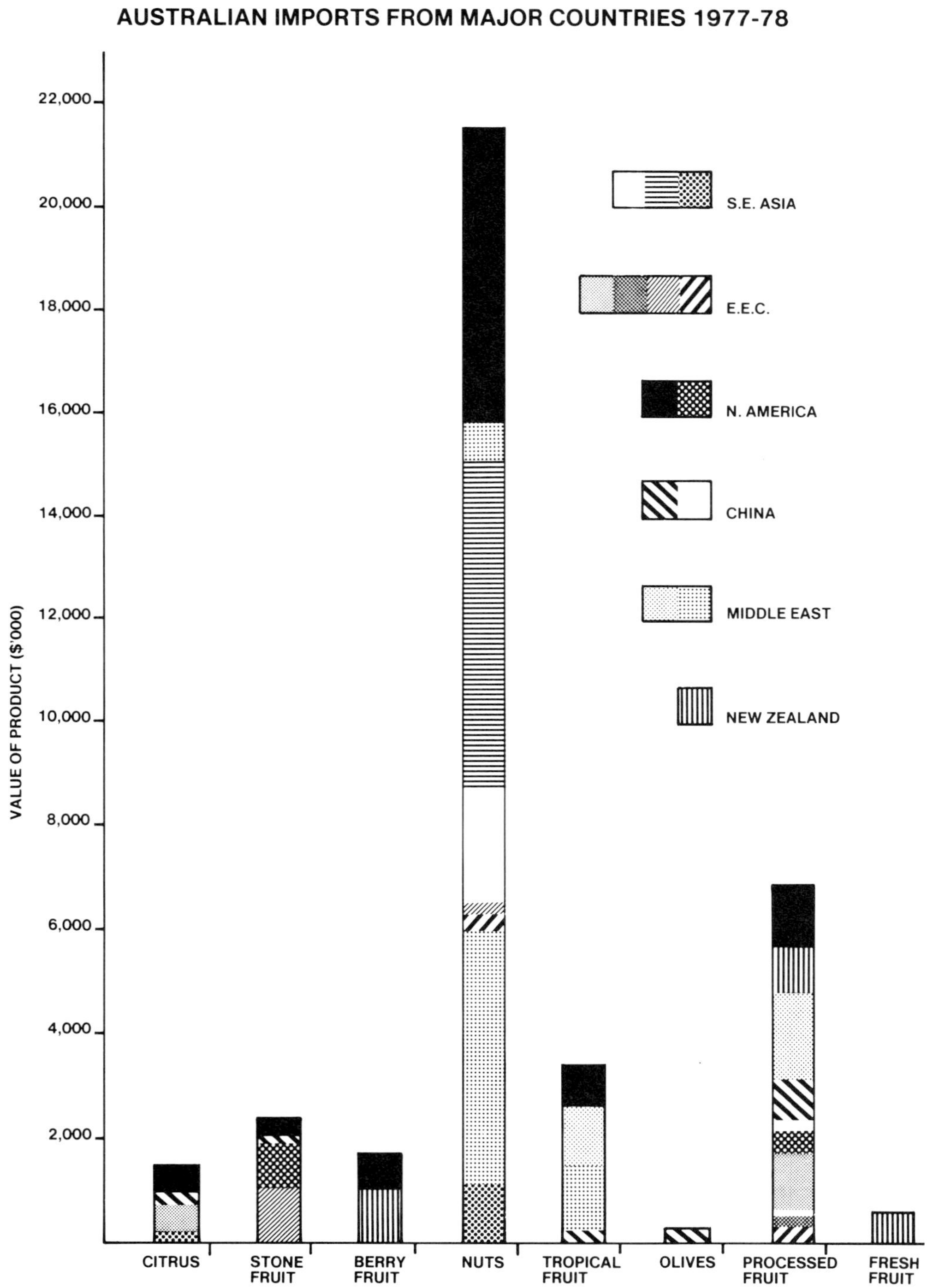

Figure 9. Australian imports for 1977-78 of fruits (excluding vine fruits) from countries for which imports of fruit and vegetables were valued at more than $10,000,000.

Chapter 2

FACTORS IMPORTANT IN ASSESSING THE POTENTIAL OF TROPICAL TREE FRUIT SPECIES FOR AUSTRALIAN HORTICULTURE

by B.W. CULL

This chapter discusses those factors which might be considered most important in an assessment of the potential of a tropical tree crop.

They are considered, in their likely order of priority under five main headings. Consumer Potential covers those elements that influence the demand by the public for the product, maintain their interest and might eventually establish a continuing viable market outlet. Product Marketing Characteristics evaluates those factors that control the successful transfer of quality produce from the grower to the consumer. Environmental Adaptability outlines the ability of the crop to perform under Australian conditions. Production Potential looks at those cultural areas pertinent to the economic field performance of a crop. Economic Potential summarizes the possibilities for success in the Australian market, as well as the limitations on a product's viability.

A tabulated list is also provided of the numerous, but by no means interdependent, criteria relating to each of these divisions. The criteria are allocated values to give some indication of their comparative importance. The evaluation preferably should be used as a basis for discussion, and not as a definitive analysis.

One column has been left clear for the reader to make his own evaluation of a particular crop's performance in relation to each criterion. It is not suggested however that crops should be judged by their 'tally'. There are too many varying and indefinite factors involved for such a simple analysis. However, this system of rating might be useful as a 'shorthand' summary of your evaluation, or as a reminder.

The evaluation of the crops is based on their use as food crops and not as ornamentals or drug crops. Alternative uses may be considered of additional value.

Consumer potential

Products can be sold by vigorous advertising at a cost; others sell themselves by developing a strong want in the purchaser. The latter is the better course but the maintenance of this want depends on several factors.

Quality

Quality, the all embracing term, must be such to firstly capture the want. However, quality must be consistent to maintain the want. Quality can vary throughout the year, as in pineapples. Poor quality at one period acts to deflate the market even when quality is high. Quality varies between localities and cultivars as is the present situation with papaws. Bad experiences are reflected in a general reaction and resistance to purchase the entire crop. Some discerning buyers select specific growers from whom to buy. However, this is not the basis of broad market development.

Quality can be broken into two aspects. The first relates to all those characteristics which enable the fruit to remain edible, pleasing to the eye, and saleable at the retail outlet. Such factors as disease susceptibility, ease of bruising, physiological breakdown, and dehydration, all mitigate against quality.

The second aspect is related to the joy of eating. The major factors here in order of priority are an acceptable sugar-acid ratio, acceptable flavour and pleasant texture. Major factors affecting acceptance are the presence of off-flavours, bitterness and astringency.

Buyer confidence

The maintenance of sales depends on uniform quality. Some products have acceptable eating quality in the majority of fruit produced (banana) while other fruit can be assessed in the main by appearance (Delicious apple). Variations in another group cannot be assessed by eye and must be tasted to assess quality (pineapple). Where quality control is easily achieved, buyer confidence can be maintained, assuring sales.

Market development

A short or interrupted supply is costly in market stimulation and has to compete with a drift to other products. The success of market development and its maintenance therefore depends greatly on an even supply of significant quantities of good quality fruit over an extended period.

Competition with other fruits

Competition with other products either prevents the development or limits the potential of market size. In Australia the February-March-April period is glutted with a wide range of temperate and subtropical fruit. The late winter-spring period is starved for fruit. On the export sphere the ability to place an accepted fruit commodity on the market in a gap in traditional supplies could assure an outlet. Such gaps have by far the greatest opportunity for success.

On the export market the introduction of new crops outside a region's normal cultural and dietary experience requires massive development and promotional inputs. Such input is not likely to be available for these new, comparatively small scale, tropical fruit crops.

Place in dietary chain

The potential value of a fruit crop depends greatly on its place in the dietary chain. Commercial fruits can be described as staple foods (apples, bananas); speciality foods (passionfruit, cherries); and novelty foods (jackfruit). The avocado as a savoury, entree and salad component has, as the price falls, the potential of becoming a staple food and hence an expanding market outlet. The passionfruit as a flavouring item could never achieve this.

Ease of consumption

The ease of consumption also affects adaptability of the product. Fruit can be eaten directly out of hand, out of hand with the aid of a utensil only, and prepared or cooked prior to eating. The size of the market and ease of sale tends to decrease in order of these divisions. Exceptions may occur but presently no major selling fruit line is sold only for cooking in Australia. Speciality and novelty fruits must meet the convenience requirements that rip-top canned drinks and wrapped confectionery do.

Ability to promote

A particular characteristic of a fruit may provide an outlet and means of promotion. The avocado as an entree or savoury is a good example, while kiwifruit as a decoration on deserts is another. Food value or health gimmick are possibly of initial value but such factors do not prevail unless the food has a characteristic which gives joy to eating.

Ability to sell on particular markets

Certain crops, and in fact cultivars within crops, and manner of presentation must be aligned with certain markets.

The Australian market yet unaware of many of the crops can in fact be educated to a taste. There is, however, a preference for sweetness with an acid balance rather than acid sweet. Bland favours are not relished. However, self-appointed connoisseurs of a crop tend to prefer a much stronger flavour than the general public will accept.

The export market requires fruit which suit their taste and texture requirements as well as their local cultural and religious needs. Launching a new crop or even cultivars unknown to the area, from Australia, could be very difficult and costly.

The future in export markets possibly lies in those crops to which the countries are already accustomed. Here the out-of-season market would be exploited. Under these circumstances, varieties, style of fruit, and degrees of maturity must suit their traditional requirements. The ability to obtain and grow suitable cultivars with the quality characteristics they require and place them on the market as required are the factors to be considered in establishing this outlet.

Processed product market

It appears that the major factor limiting the sale of processed products is the previous acceptance of the fresh fruit. Without a knowledge and appreciation of the fresh fruit there could be initial resistance to the processed product. Where the crop has attributes making it suitable for processing, the use of dual purposes or individual fresh fruit and processing lines, are needed to establish simultaneous markets.

Product marketing characteristics

Marketing could be defined as the steps which follow harvest — preparing, holding and transporting the product to reach the point of sale.

Here we must look at two individual streams, fresh fruit and processed fruit. There is, however, an overlapping in needs for both outlets.

FRESH FRUIT

Fruit life

The life of fresh fruit depends on the maturity (edible stage) status of the fruit at harvest. Fruit can be either immature at harvest and ripen with time, while others must reach full maturity before harvest. The former type has considerable advantages.

The life of the fruit following harvest (transit and shelf) must be sufficient for it to reach the consumer and then remain usable for several days. Preference would be given to fruit which are in an edible state at the market and maintain this characteristic for an extended period (apple) rather than those which have only an acute peak of acceptability (banana).

Fruit life may be extended by either cool storage or controlled atmosphere techniques. Such manipulation can mean the success or otherwise of a crop, although the cost is significant (kiwifruit).

Fruit storage

The ability to store fruit enhances its usefulness for fresh and processing outlets by facilitating the development and control of transport techniques and general holding and redistribution. The process of reaching distant markets and retail points can capture sales. Storage also allows a controlled and extended supply to the market place which will maintain the market. Similarly the processor can control the rate of flow or fruit through the plant, thus reducing the capacity needs in equipment and staff.

Cool storage presently dominates most aproaches to crop storage. Controlled Atmosphere (C.A.) techniques appear more involved and the possibility of having detailed information on undeveloped crops is unlikely.

Packaging

The case, type and cost of packaging relates to the shape and susceptibility to damage of the product. Those having uniform and preferably rounded shape fit automated sizing, transfer and packaging. Fruit with a capacity to support a package (orange) obviously is cheaper to carton.

Transport

Fruit value to weight ratio controls the costs of packaging and transport, and becomes more significant the further away markets are, and the more perishable the product. The more perishable the product the faster and more costly the mode of transport needed. The costs of transport increase abruptly as the need for controlled temperature and atmosphere storage becomes necessary.

Trade barriers

Trade barriers associated with quarantine exist both interstate and overseas. The susceptibility of fruit limits their value. Disinfestation and treatment techniques are available for some crops. Such techniques are costly and also impede the process of crop assemblage and forwarding. The ability to assess this area depends on the availability of past research and development input.

PROCESSED FRUIT

Type of supply

The development of a processing industry in the main cannot grow from the chance excess production or discards of a fresh fruit industry. As crop production becomes large this rule changes in part. However, guaranteed supply over an extended period, within limits, is a must. A further rider is that some fresh fruit varieties do not meet the needs of a processor and special cultivars are required.

Type of processing

Suitability of a crop to either drying, freezing or canning alters its value in the development of processing. Small freezing and drying plants, provided they can get the continuity of supply from one or more crops, can be viable. Canning of goods in remote areas has the major problem of shifting tinplate in two directions. In addition, the general canned fruit market is somewhat depressed. The production of juices is usually a simple and low capital cost process. However, the market is presently limited and highly competitive.

Processing plant siting

In newer crops one can see a limited opening for such processing plants. Continuity of supply is possibly the major limiting factor. Alternatives are, to find a range of crops to give the supply, or transport the product to a processor with an already established throughput. The former requires the simultaneous establishment of the range of suitable crops at a limited distance from the plant. The possibility of this happening is remote. With established fruit processing plants being presently centred no further north than close to Brisbane, a processing crop established in far flung areas could require characteristics in the fruit which would enable it to be economically transported over a long distance. Another alternative is to have some simple processing step carried out on the grower's property.

Environmental adaptability

We accept that environment encompasses both soil and climate and that within what is classed as a tropical environment both factors range widely due to the original geology and present land relief. Adaptability refers to the crop's ability to exploit a wide range of these situations. Exploiting would mean consistently producing commercial crops and not merely surviving. Advantage can then be made of significant land areas for production and the variations in the related cropping times to exploit markets.

Location of production areas is often governed by the historical location of man and the services which serve him and his horticultural activities. The wider the adaptability the greater the chance of a crop being exploited. Where limited genetic diversity restricts fruit harvesting periods to weeks (mango and lychee) fresh market development through extended supply can only be achieved by having production areas in differing environments or by costly storage systems.

The criteria which are listed are considered those most suitable to define the adaptability of a horticulturai tree crop. They also type the crop sufficiently so that its location can be defined with respect to the environment.

Consideration is given to temperature, water, wind, soil, land use and cultural practice, and environmental adaptability.

Production potential

Production performance is related to crop size and regularity of such cropping, year by year. This is further coloured by the ease or ability to bring commercial areas to this status within an economical time scale. Performance can be jeopardized by the susceptibility of the crop to pests and diseases and weather conditions. A further controlling factor of performance is the ease of harvesting.

Bearing capacity

The ability to bring sizeable plantings into production quickly spells potential financial success for any crop. Ease and cost of producing trees and their early development are important. The size of the early crop and the ultimate crop all govern its potential success. The actual production life of the tree has a bearing on return to investment. Life can be measured as either when the plant is replaced or when the plant has to be rejuvenated or skeletonized to bring it back to a controllable unit.

Tree and crop management

The ease and cost of manipulating the size, structure and shape of a tree is important with respect to the structural stability of the tree and the economics of cultural procedures and harvesting. The ability to control yield limitations or manipulate trees to yield more in a given time to improve the ease of harvesting are factors in favour of a crop. Factors which may be manipulated are tree growth and vigour, flowering and fruit set. This ability does not totally negate the defect as there can be considerable cost involved.

Usually in evaluating crops, the ability to assess these manipulators depends greatly on the accrued technical background which is related to previous input.

Crop loss

The incidence of pests and disease and also environmental damage in a specific region will have the ultimate control of crop loss. This is difficult to measure before plantings are made. What can be evaluated is the potential susceptibility of the crop to likely problems.

Ease of harvesting
This is a function of the cost-benefit relationships of time spent accumulating a certain value of product. This is usually controlled in part by the size of the fruit unit, its resistance to damage, the ease of finding it in the tree and its dislodgement, and the height of the fruit above ground level. These factors are magnified dramatically by the number of times you have to return to one tree to complete the harvest cycle.

Crops may be divided into: those that ripen only after removal from the tree (avocado); those which mature and will hang, retaining their quality (orange); those which mature fairly evenly on a tree but must be harvested immediately (peach); and those which require to be harvested immediately when mature but which on a given tree, mature over a period of time (custard apple). The first one obviously has great benefits both in the harvesting and marketing spheres. The second, like the first, also assists management and allows time for solving problems arising in the harvesting, marketing and processing chain. The third type is subject to acute loss by misfortune due to environmental and marketing factors at fruit maturity. The fourth tends to spread the risks in this area. However, the cost of repeated harvesting tends to limit the value of the crop.

Economic potential

The future and continued success of any crop depends on the financial return for investment in labour and capital. In addition certain economic factors control the type of person, related to his access to capital, entering the potential industry.

Likely grower return
Selling infers returns to growers which will allow continued marketing. Likely returns per kilogram of fruit may in the long term be equated against similar products in the dietary chain today (lychee may be compared with cherries). Many of the overhead and cultural costs of production are similar for tree crops. Harvesting, packaging and marketing costs can vary significantly. Where the latter costs are equated to be similar, then potential weight of marketable fruit should give a fair estimate for comparison of the likely financial viability of a new crop.

Capital investment
The capital investment required depends on two major factors. The first is the net return per unit area. Where the returns are high (avocado) the actual initial investment in tree crops need not be high. Where the returns are lower the scale of investment needs to be far greater to cover larger areas (macadamia).

The second controlling factor is the years to break even between costs and returns. The longer the period the greater the initial capital input. Large capital usually involves investors of some degree who are more likely to place their money in industries with a performance record than in potential new crops. Therefore new crops with a lower initial capital input are more likely to make a start than those requiring intensive capital investment.

Integration with established crop
If a crop can be generally coupled with an established industry or in fact progressively replace it, then capital invested in machinery, packing and storage facilities do not require duplication. This would also apply to cultural and management procedures making the adoption of the crop by established growers far easier.

Other uses

The tropical tree fruit crops have additional roles in many cases. Such roles may act as additional income to the production of fruit or as separate activities. Drug production, industrial chemical products, timber, the sale of ornamental cultivars and suitability for non-commercial backyard production are worth noting.

Evaluation of the potential of tropical tree crops

Using the understanding described in the section on factors important in assessing the potential of fruit species, the relevant criteria have been laid out in a structural fashion for easier use.

Against each criterion a nominated criterion value has been given. This value indicates the relevant importance, in my opinion, of each criterion. These values can be the subject of debate and possible change. The values are based on the following verbal value judgements.

Vital	Highly desirable	Desirable	Useful	Inconsequential
10	8	6	4	2

The use of this list of criteria in evaluating a crop may be considered in two ways. The criteria could be used as a check list and by working systematically through them, most important factors related to the potential of a crop are reviewed. The strengths, weaknesses and areas of lack of knowledge are revealed. Research and development areas should follow from this assessment.

Those wishing to evaluate marginally may use the following rating system to record the performance of a given crop against each criterion. It is suggested that by using the rating you are forced to make an active decision about each criterion and in recording it you have the opportunity to review the situation at a later date. It is not recommended that candidate crops be compared by a numerical accumulation scheme. The suggested performance ratings are:

High	Medium	Low	Nil
10	5	2	0

These are a guide and ratings may be allocated between 0 and 10.

Consumer potential	Criterion value	Performance rating
Quality		
Ability to maintain good quality throughout production period	10	
Ability of all cultivars used to maintain a good and similar quality standard	10	
Acceptability of visual quality characteristics at the market place	10	
Acceptability of the sugar-acid ratio of fruit	10	
Acceptability of unique flavour of fruit	10	
Acceptability of texture of fruit flesh	8	
Buyer confidence		
The relationship between visual characteristics and eating quality	10	
The relationship between texture (softness) and eating quality	6	
Market development		
Ability to maintain a supply of significant quantity of good quality fruit on the market (high: 4 months; low: 6 weeks)	10	
Competition with other fruit		
Ability to place fruit on the Australian market in autumn-early winter	6	
Ability to place fruit on the Australian market in late winter-spring	10	
Ability to place fruit in a gap in traditional supply overseas	10	

	Criterion value	Performance rating
Place in dietary chain		
Characteristics of fruit are such that they fulfil a significant role in the dietary chain. High = staple food; Med. = speciality; Low = novelty	8	
Ease of consumption		
Ability to be consumed easily	8	
Ability to promote		
Ability to promote on some specific use characteristic	8	
Ability to promote on some food value	6	
Ability sell on particular markets		
Possible degree of acceptance by a wide range of the Australian public or ethnic group	10	
Ability to grow cultivars which are now marketed overseas for that market	8	
The availability of known markets with gaps or opportunities which do not require extensive development	8	
Ability to sell at an acceptable price	8	
Ability to replace imports	10	
Processed product market		
Ability to exploit the fresh fruit market simultaneously	10	
Ability to be marketed and used without competition from a similar product from another established crop	8	
Product Marketing Characteristics		
FRESH FRUIT		
Fruit life		
The extent of fruit life after harvest without cooling — high: >14 days; low: <7 days	8	
The extent of fruit life after harvest with cooling — high: >60 days; low: <14 days	8	
Span over which fruit becomes edible following harvest — high: for whole period low: <3 days	6	
Fruit Storage		
The ability to extend fruit life by cooling	8	
The ability to extend fruit life by C.A. storage	6	
The ability to hold ripe fruit in a household refrigerator	8	
Packaging		
Ability of fruit to withstand mechanical damage	8	
Ability of fruit to easily fit automated sizing and packaging equipment	6	
Ability of fruit to support the package	4	
Transport		
Value by weight ratio (strawberry — high; bananas — low)	6	
Ability of product to be marketed without sophisticated packaging	6	
Ability of product to be marketed without storage requirements	8	
Proximity to market	4	
Trade barriers		
Ability of fruit to resist attack of Queensland fruit fly and other fruit flies	8	
Ability of product to be marketed without pesticide treatment for rots and infestation	8	
Ability of fruit to withstand without damage ethylene dibromide fumigation or gamma irradiation	10	
PROCESSED FRUIT		
Type of supply		
The potential for processing is such to stimulate separate production for processing	8	
The ability to use fresh fruit varieties for processing	8	

	Criterion value	Performance rating
Type of processing		
Ability of the product to be frozen	8	
Ability of the product to be dried	6	
Ability of the product to be canned	4	
Ability of product to be juiced	6	
Processing plant siting		
Ability of the crop to supply a processor continuously with commercial quantities for more than 4 consecutive months	10	
Ability of product to be transported in bulk 500 km	8	
Ability of product to be stored and transported in a dried form	6	
Ability of product to be simply processed for storage and then transport on the property	6	
Environmental adaptability		
Temperature		
Crop suited to:		
Tropical climate only	6	
Tropical and subtropical	8	
Subtropical	8	
Subtropical and semi-temperate	8	
Temperature stress:		
Ability to withstand 3 consecutive days at 38°C	6	
Ability to perform with extensive periods 32°C	8	
Ability to perform with extended minimum screen temperature of 0°C to 10°C	8	
Ability to survive minimum screen temperature of –4°C to 0.0°C	4	
Water		
Rainfall pattern adaption of crop:		
Suited to heavy wet summer — moist winter pattern	8	
Suited to heavy wet summer — dry winter pattern	8	
Suited to moist summer — dry winter pattern	6	
Suited to dry summer — moist winter pattern	4	
Poor water quality adaption:		
Ability to use irrigation waters 150 to 300 ppm chloride	8	
Ability to use irrigation waters 300 ppm chloride	8	
Response to humidity:		
Ability to withstand high humidity at:		
flowering	8	
carrying crop	8	
Ability to withstand low humidity at:		
flowering	8	
carrying crop	8	
Wind		
Response to wind extremes:		
Ability to perform in prevailing winds 15 knots:		
during vegetative growth	6	
during flowering	6	
Ability to withstand dry, warm to hot winds:		
during flowering	8	
at carrying crop	8	
Ability to withstand structural damage in cyclonic and storm winds	8	
Soil		
Adaptability to soil types:		
Range of adaptability to internal drainage	8	
Range of adaptability to soil pH	4	
Range of adaptability to soil fertility	4	
Land use and cultural practice		
Resistance to erosion risk	8	
Ability to use wide variations in topography	6	
Ability to use wide variation in orchard surface	6	

	Criterion value	Performance rating
Environmental adaptability		
Range of genetic environmental adaptability exhibited by cultivars	8	
Production potential		
Bearing capacity		
Ease and efficiency of multiplication of clonal material	6	
Years to commercial harvesting (5 years: high; 10 years: low)	8	
Ability to bear crops of an economic size	10	
Reliability of annual cropping	10	
Length of commercial economic harvesting life (20 years: high; 10 years: low)	8	
Tree and crop management		
Ease of cultural management	8	
Ability to control tree vigour	6	
Ability of plant to flower sufficiently and consistently	10	
Ability of plant to carry set fruit through to harvest	10	
Extent of reported technical knowledge on crop performance	6	
Crop loss		
Degree of resistance to known important insect pests	8	
Degree of resistance to known important diseases	8	
Degree of resistance to environmental stress during growing period	8	
Degree of resistance to environmental stress during cropping period	8	
Ease of harvest		
Ability to remove fruit on one tree at one harvest	8	
Ability to hold fruit on tree after reaching an acceptable market stage (3 weeks: high; 1 week: low)	8	
Ease of harvest of fruit from ground in mature tree	6	
Ability to harvest fruit from external surface of tree	6	
Predisposition to mechanical harvesting of crop	8	
Resistance of fruit to mechanical damage	8	
Economic potential		
Likely grower return		
Likely financial return based on comparison with another crop — nominate crop	10	
Capital investment		
Ability to establish a viable production unit on low investment in land (2 ha high; 6 ha med.; 20 ha low)	6	
Ability to establish a viable production unit on common standard farm machinery	6	
Ability to establish a viable production unit without large investment in harvesting, processing and storage equipment	8	
Ability to reach break even point between costs and returns early (5 years: high; 8 years: med.; 10 years: low)	8	
Integration with established crop		
Ability to integrate with an established crop industry in Australia — nominate crop	8	
Other uses		
Nominate uses		
Ornamental		
Chemical		
Drug		
Timber		
Backyard		

Chapter 3

ANACARDIACEAE

CASHEW

by J.D. STURTZ

Anacardium occidentale L.

Botany

The cashew tree is a spreading evergreen with rough bark, and reaches a height of 12 m and a diameter of 20 m. Leaves are alternate, simple, obovate, 6 to 20 cm long, and 4 to 15 cm broad. The inflorescence is a conical, terminal panicle to 20 cm long, bearing both male and hermaphrodite flowers. Individual flowers have 5 green sepals 5 mm long, and 5 greenish cream, turning red petals 1 cm long. Male flowers generally have 9 short stamens with 1 longer stamen projecting above the corolla. Hermaphrodite flowers are similar except that a 12 mm simple style also projects above the corolla. The ovary has a single ovule. The fruit is a green, turning greyish brown, kidney-shaped nut (achene) about 3 cm long and 2.5 cm broad with a hard honeycombed pericarp partly embedded in an enlarged (10 to 20 cm long, 4 to 8 cm broad), red to yellow, pear-shaped, juicy, peduncle ('apple'). Cashews are cross-pollinated.

Climatic requirements

Cashews are drought and salt-spray resistant, intolerant of frost and extreme winds, and grow in tropical climates with 400 to 4000 mm rainfall (optimal rainfall 750 to 1300 mm, temperature 15 to 32°C). Too much rain, or humidities below 50% at flowering and fruit set, are undesirable.

Growth is excellent on the wet tropical coast of Queensland but high rainfall and humidity increase pests, diseases, and nut damage. Drier areas are considered more suitable.

Growth is satisfactory between Katherine and the coast in the Northern Territory where monsoonal rainfall (960 to 1500 mm) generally allows completion of harvesting before 'the wet'. Cashews also grow in the Kimberley region of Western Australia.

Soil requirements

Cashews prefer sandy soils of medium fertility, but also grow on well-drained clays and shallow, stony and lateritic soils. Brackish, waterlogged or alkaline soils are considered unsuitable.

In the Northern Territory, trees have grown better on deeper sands, sandy loams and earths than on shallow laterites. Sandy soils at Kununurra and Broome, beach sands in Queensland and the Northern Territory, and undisturbed soils at Weipa have proved satisfactory.

Genetic diversity

Variations have been recorded in disease and pest resistance, tree habit, flower production and shedding, flowering and fruiting times, and 'apple' colour. Dwarf types have been reported in Amazonia. Only recently have superior lines been selected in East Africa and India.

Tree and fruit characteristics of best cashew strains at Kamerunga Horticultural Research Station, Redlynch.

Prepared by A.J. Wait

Selection and tree no.	Origin	Tree habit	Canopy	Height (m)	Tree age 1981	Av. yield last 4 yrs (kg)	Nut wt (g)	S.G.	% Kernel	% CNSL	Apple shape	Apple colour	Apple size (g)
Kam. (2)	Local	spreading	medium	10	20+	9.58	8.3	0.98	23.5	22.6	blocky	red (3)	90
Kam. (6)	Local	spreading	medium	10	20+	15.42	7.2	—	—	—	blocky	red (3)	90
MS252 (70)	Kenya	spreading	open	6	8	6.50	5.4	0.89	N/A	—	pear	yellow (1)	60
A81 (75)	Kenya	spreading	medium	5	8	5.88	4.3	1.06	26.4	17.3	blocky	red (5)	50
MS90 (81)	Kenya	erect	open	7	8	2.26	7.7	—	—	—	pear	yellow (2)	80
AB9 (91)	Kenya	spreading	dense	7	8	2.35	4.7	—	—	—	blocky	red (4)	60
MS228 (97)	Kenya	spreading	medium	5	7	1.76	5.6	1.05	24.4	20.0	pear	red (3)	80
ATA58 (109)	Tanzania	spreading	medium	7	7	2.72	6.2	0.99	24.6	16.5	blocky	red (3)	100
AC4 (127)	Sri Lanka	spreading	medium	6	7	2.66	7.4	0.97	28.2	20.4	pear	red (4)	60
AC14 (160)	Sri Lanka	erect	medium	8	7	3.80	5.8	1.03	29.5	16.8	pear	yellow (1)	70

Definition:

Tree habit — spreading, erect.
Canopy — open, medium, dense
Apple shape — blocky (height $1–1^1/_2$ times diameter); pear (height $1^1/_2$–2 times diameter); elongate (height 2–3 times diameter).
Apple colour — (1) fully yellow; (2) yellow, small amount of red; (3) mid-yellow/red (fawny red); (4) predominantly red; (5) fully red.

Early introductions into the Northern Territory came from Java in 1941–42 and their seedlings show some variation.

In the early 1970s the Queensland Department of Primary Industries imported seeds from Asia, Africa and New Guinea and is comparing these with Queensland and Northern Territory seedlings. Yield per tree, specific gravity of nuts, time and duration of fruiting, and general growth and fruit characteristics are being recorded. Some trees have better yields and larger nuts. Most are biennial bearing (2 to 3 fold variation), exceptions being A81 and MS252, while nuts from the AZA2 and C5 have good weathering ability. Anthracnose resistance is superior in trees with orange, non-elongated 'apples'. The best lines are Kamerunga, MS252, MS90, A81, AC14 and ATA58.

Production season and time of cropping

Flowering occurs during the dry season with the fruit becoming visible within a week. The fruit reaches its full size in about 4 weeks, followed by 'apple' swelling and nut ripening by about 7 weeks, coinciding with the late dry/early wet season.

In Queensland flowering commences in late winter and early spring with fruit setting by October, extending over 6 to 10 weeks. The main cropping period is November-December continuing into January, when rain and humidity can cause germination or deterioration of nuts before harvesting.

In the Northern Territory flowering commences in early June with occasional out-of-season flowering in December. Fruit drop starts in mid-August, peaks in September and October, and is largely completed by November, generally before the onset of regular rain.

Yield potential

Yields overseas vary from 0.5 to 50 kg per tree and from 100 to 2000+ kg per hectare.

Each panicle bears from 100 to 1600 flowers with a ratio of male to hermaphrodite of 6 to 30:1 (Northern Territory 6 to 83:1). From 1 to 24% of hermaphrodite flowers produce mature nuts. Fruit per panicle may be 1 to 6 (Northern Territory 0 to 6). Insect and disease damage, ineffective pollination, and adverse climatic conditions reduce effective fruit set.

Australian nut yields (kg per tree)

Site	Age (years)					
	3	4	5–6	7–8	10–12	Mature
Kamerunga (Qld)			0.17–2.5	1–8		8–20
Berrimah (N.T.)			0.30–7.0			
Garden Point (N.T.)				3–5	12	1–24
Daly River (N.T.)		5				
Millingimbi (N.T.)	1		2.0–4.0			

Bearing commences about year 3, is earlier in vegetatively propagated trees, and reaches a peak about year 10. The oldest trees at Garden Point are almost 40 years old and still prolific.

Yields are not particularly high, indicating a need for relatively large plantations.

Market acceptability

Cashews are readily accepted as a savoury or dessert nut, particularly in the more affluent countries. Being a luxury food, consumption and prices reflect general economic conditions in affluent countries, seasonal supplies and prices of competing nuts. Whole kernels are preferred, broken pieces and scorched wholes or brokens being downgraded.

Cashew nut shell liquid (CNSL), a drying oil, is used in the manufacture of paints, resins and synthetic fibres.

The 'apple' is consumed as fresh fruit, pickles, jellies, jams, chutneys or wine in producing countries.

Export potential

World production and consumption has increased steadily over the last 25 years. Major producing countries are India, Mozambique, Tanzania, Kenya and Brazil, and major consumers are North America and Europe. India and African countries are attempting to increase both area and yields.

Cultural details

Propagation

Direct seeding is most common. Propagation by marcotting, budding, grafting and cuttings is possible, but is considered expensive, unreliable, and complex.

In the Northern Territory we transplanted seedlings, finding high density seeds, early planting and larger bags gave best results.

Management

Land must be cultivated as direct sowing into pasture is unsuccessful. Inter-row cropping is common overseas, and in Australia could feature a pasture legume or an annual crop.

Weeds must be controlled at establishment to reduce insect damage and competition, and later to facilitate harvesting.

Spacing varies with soil type, rainfall and inter-cropping from 3 x 3 to 10 x 10 m initially, up to 20 x 20 m with large mature tress (1000, 100 and 25 trees per hectare). Wide rows and close initial intra-row spacing may facilitate machinery operations and maximize early yields.

Pruning is unnecessary. Lower branches can be removed to facilitate harvesting and weed control.

Irrigation and fertilization

Cashews are seldom irrigated or fertilized, although responses have been recorded.

Pests and diseases

Cashews are relatively resistant to pests and diseases with tea mosquito *(Helopeltis)* and anthracnose *(Colletotrichum)* being most serious overseas.

In Australia bananaspotting bug *(Amblypelta),* black beetle *(Rhyparida),* red banded thrips *(Selenothrips),* termites, and possibly stem borers *(Bombatelia)* and anthracnose have caused most damage. Incidence depends on site and seasonal conditions. Strain resistance has been observed in Queensland.

Harvesting methods

Overseas, nuts are hand harvested. In Australia machine harvesting, before rain damage, would be required using United States harvesters or by developing similar machines.

Marketing characteristics

Raw nuts are sold according to size and quality, and consist of 20 to 30% kernel and 70 to 80% shell containing 15 to 30% CNSL. Kernels are sold as: whole whites (grading W210 to W500 — number of kernels per 500g); scorched wholes, dessert wholes, white pieces, scorched pieces; dessert pieces.

Average quality of Northern Territory nuts processed in India in 1965 was less than Indian nuts, although some lots were as good, and comparable with, Mozambiquan.

Storage and transport characteristics

Raw nuts may be dried and stored for up to 2 years. Processed kernels are stored in evacuated or carbon dioxide treated, 12 kg tins for 1 year or longer.

CNSL is transported in 200 L drums.

Processing characteristics

Until recently most nuts were manually processed in India by: cleaning; humidifying; grading; roasting to cook the kernel and liberate CNSL; decortication by hand; peeling; rehumidifying.

Mechanical decortication involving pressure or centrifugal cracking, cutting or sawing, has been developed. One plant with a capacity of 1250 tonnes per annum costs about $260,000 (1980) including grader, roaster/decorticator, drier, packer and laboratory equipment with a mechanical peeler as an optional extra.

An Australian cashew industry would need to install mechanical processing to supply the local market direct.

Acknowledgement

Jim Wait and Brian Watson of Queensland Department of Primary Industries, assisted greatly by providing information on cashew production in Queensland.

Further Reading

Ohler, J.G. (1979), Cashew. Communication 71, Royal Tropical Institute, Amsterdam.

MANGO
Mangifera indica L.

by A.W. WHILEY

The mango is thought to originate from the Indian-Burmese border region and has been under cultivation in India in excess of 4000 years. It is widespread throughout the tropics and subtropics and is consumed fresh in larger quantities than any other fruit. It is the most important emerging fruit crop in tropical America.

Botany

All cultivars of mango belong to the single species *Mangifera indica* L. The genus *Mangifera,* however, contains 41 species distributed from India and Sri Lanka to the Philippines (Mukherjee 1971) with many producing edible fruits.

The mango is a deep rooted, dome shaped, evergreen tree growing up to 40 m tall. The leaves are simple, entire, leathery, 8 to 40 x 2 to 10 cm, narrowly elliptic or lanceolate and produced in flushes; the young leaves are usually reddish, later turning dark shiny green and retained for approximately 1 year.

The inflorescence is a widely branched terminal panicle 10 to 60 cm long with 1000 to 6000 flowers. It is polygamous with male and hermaphrodite flowers in the same inflorescence at a ratio of 1 to 36% of the latter, higher in some cultivars.

Both flower types are about 6 mm in diameter when open. They usually have 5 green, pubescent sepals, 5 off-white, pink or purplish petals, and a 5-sectioned, yellow-green nectary disc. Male flowers usually have 1 functional and 4 abortive stamens. Hermaphrodite flowers contain, in addition, a shiny, green, globose superior ovary and a style with a single stigma.

The fruit is a fleshy drupe, varying in size from 2.5 to 30 cm long. The shape varies from rounded to ovoid-oblong and is sometimes laterally compressed. The fruit are varying tones of green, yellow and red.

The seed is enclosed in a stony endocarp, varying in size with two fleshy cotyledons. Some seeds are monoembryonic with a zygotic embryo only; others are polyembryonic with 2 to 12 embryos in which apomictic embryos are produced from the epidermal cells of the nucellus and in which the zygotic embryo may or may not be suppressed. Most Indian cultivars are monoembryonic with the evolution of polyembryonic lines in Malaysia and the Philippines regions.

Climatic requirements

The mango grows under a wide range of climatic conditions but its profitable cultivation is limited by temperature and precipitation patterns. It is reported that growth is minimal at 4 to 10°C with optimum growth and productivity around 24 to 27°C. In areas of low humidity fruit sunburns when temperatures exceed 46°C and vegetative growth ceases.

The susceptibility of the mango to frost varies with its age and stage of growth. Young plants in active growth are killed at temperatures of –1 to –2°C while older trees with low growth activity usually escape damage at this range, provided the duration of low temperature is not prolonged. Some varieties are more sensitive to frost injury than others.

Low temperatures at flowering can cause increased numbers of parthenocarpic fruit. Lakshminarayana and Aguilar (1975) report 73% parthenocarpy with the variety Haden when ambient temperatures fell to 2°C during flowering, but only 53% when temperatures were maintained at not less than 10°C. Kadman and Oppenheimer (1978) report parthenocarpy with Haden in Israel with low temperatures during flowering. Similar results have been observed with Carabau in south-east Queensland. Some cultivars appear more sensitive to temperature than others.

The mango grows over a wide range of moisture regimes being quite drought tolerant as well as capable of standing up to heavy rainfall. However, productivity of the tree is related to rainfall distribution, with flowering and fruit set requiring a dry season. Trees tend to remain vegetative in tropical regions receiving regular heavy rainfall. Beal and Newman (1978) investigating climatic influences on mango production patterns at Bowen, Queensland, obtained correlations between annual production and both summer rainfall and annual rainfall. They also significantly correlated the number of cold days in May to July (pre-flowering period) to annual production.

Mango trees in the tropics grow from sea level to 1200 m, but seldom bear at the higher elevation. In South Africa, commercial viability ceases at 600m. Altitude has a pronounced effect on time of flowering with every 120 m increase in height delaying time of flowering by approximately 4 days. Similarly for each degree of latitude south or north of the tropics flower initiation is delayed by approximately 4 days.

Soil requirements

Mango trees will grow on a wide range of soils but prefer deep well-drained sands to loams with the water-table no higher than 2.5 m. The trees can withstand occasional flooding and are probably one of the most tolerant of the subtropical–tropical fruits to 'wet feet'. High productivity can be obtained on light, low-fertility soils, possibly due to stress promoting heavy flowering and fruit set while heavy fertile soils with good water retaining capacity often produce vigorous vegetative growth at the expense of fruitfulness.

The mango will tolerate calcareous soils with a pH to 8 but does best at a range of 5.5 to 7. It does not thrive in very acidic soils with a pH lower than 5.0

Genetic diversity

Immense diversity of type exists in this species which is best typified by the 1000 named varieties in India. No doubt this is a reflection of both the monoembryonic character and the length of cultivation of the mango in that country. However, in the commercial sense the industry is based on 30 varieties in India of which 12 could be considered of major importance (Chadha, personal communication). Like most tree fruit crops all major varieties have risen as chance seedlings.

Meulen (1971) estimates the number of mango cultivars grown commercially throughout the world exceeds 100 and suggests narrow environmental adaptation of each variety as the reason for this large number.

Within the recognized mango varietal population is a wide range of:

- Tree habit: ranges from the compact, low vigour to the fast growing, large trees.
- Fruit size and shape: from 0.2 to 11 kg and rounded to ovoid-oblong.
- External fruit colour: red, pink, blue-green, green, yellow.
- Internal flesh quality: ranges from heavily fibred (undesirable) to slightly fibred.
- Flavour: Meulen (1971) describes four flavour types: (i) pronounced turpentine-like taste; (ii) 'wild' pungent flavour sometimes with carroty taste; (iii) fine flavour and very sweet without any acid; (iv) delightful mango flavour blended with acid being most acceptable to the European palate.

Srivastava *et al.* (1978) are selecting dwarfing rootstocks based on stomatal leaf count and have established a relationship for selection while Kadman *et al.* (1978) found sufficient genetic diversity in seedling populations to select for salinity tolerance.

Production season and time of cropping

Some polyembryonic mango seedlings start bearing at 5 to 6 years of age but can take up to 10 years (Meulen 1971; Valmayor 1968). Some grafted monoembryonic

Some mango cultivars

Variety	Fruit shape	Skin colour	Maturity time	Fibre	Eating quality	Anthracnose resistance	Bacterial spot resistance
Kensington Pride	ovate to slightly oblong	yellow with pink blush	mid season	moderate	excellent	low	moderate
Common	oblong to elongated	yellow	early	high	fair	moderate	high
Banana–1	long oblong	yellow	early	low	very good	low	moderate
Glenn	oblong–ovate	orange/yellow pink blush	early	low	very good	moderate	moderate
Haden	ovate	yellow with brilliant red blush	mid season	moderate	very good	low	low
Zill	oval	yellow with dark red blush	mid season	low	very good	moderate	moderate
Carrie	ovate to slightly oblong	yellow	mid season	low	excellent	low	low
Irwin	ovate to slightly oblong	bright red	mid season	low	excellent	low	very low
Kent	ovate to slightly oblong	red blushed	late	low	excellent	low	very low
Tommy Atkins	ovate to slightly oblong	dark red	mid season	moderate	good	moderate	moderate
*Dashehari	oblong to elliptical	yellow	mid season	very low	excellent	low	—
*Alphonso	ovate–oblique	yellow	mid season	low	excellent	low	—
*Bombay green	ovate	yellow-green	early	moderate	very good	low	—
*Langra	oblongish	lime-green	mid season	moderate	very good	low	—

* Important Indian varieties. Maturity times refer to data from India.

varieties bear in their third or fourth year after planting out with most in production by their sixth year (Beal 1981).

Flower initiation usually occurs during autumn with flowering occurring from midwinter to early spring depending on the variety and temperature. The time of development after fertilization to maturity of fruit is 2 to 5 months. Some cultivars in the tropics set a few fruits throughout the year in addition to their main cropping period.

The harvesting of mangoes in India extends over a period of 7 months (Chadha 1975); in South Africa it extends for around 4½ months (Meulen 1971); and in Florida about 4 months (Malo 1977).

The period of cropping can be greatly extended by the choice of early, mid and late season cultivars in any one district. In Australia, the mango industry is based predominantly on one variety with the major production centre in the dry topics. This has led to a short mango season of approximately 6 weeks. Beal (1981) concludes that 'utilization of selected varieties from the early and late groups together with Kensington Pride could extend the mango production season in the Bowen district to at least 12 to 14 weeks'.

The potential for a greater extension of the season in Australia exists by use of environmental factors. The tree will grow and crop from the northern tip of Australia to coastal areas at approximately latitude 28°S. Disease problems encountered with Kensington Pride in the southern areas may be overcome by use of tolerant varieties such as Tommy Atkins, Florigon and Carabau.

Yield potential

Biennial or irregular bearing is well known with the mango (Popenoe 1920, Meulen 1971, Chaudhri 1976). While these patterns can be partially attributed to climatic influences (see Climatic Requirements), the cultivar can have an overriding effect. The flowering intensity of cultivars varies markedly as does the sex expression of the flowers. Meulen (1971) states that perfect flowers account for 2 to 20% depending on cultivar, while Purseglove (1968) gives a range of 1 to 36%. The other major contributing factor to low yield, which may be either climatically or nutritionally controlled, is the failure of mature fruit-bearing shoots to differentiate flower buds for the following crop after harvest (Rao *et al.* 1978).

Few production figures are available in the literature. However mature trees of select cultivars may produce up to 500 kg per tree.

Market acceptability

The mango is claimed to be the most important fruit of the tropics. Once known in India as the 'Fruit of Kings', then the 'King of Fruits' it is now the 'Fruit of the Masses', (Chadha, private communications).

While the external appearance of the fruit can be considered attractive — smooth skin green-yellow with pink blush through to red-crimson — the internal quality is a decisive factor on acceptability. Undesirable features include coarse fibrous strands through the flesh and the 'wild' pungent and turpentine flavours of some varieties. The latter is acceptable to Asian consumers but not to the European palate.

However, many excellent varieties exist which should find a place in the Australian markets.

Export potential

Outside of the tropics and subtropics the mango is little known. Export of fresh fruit from Australia would have to be aimed at high priced western markets to off-set freight costs. One of the greatest problems is the landing of fruit onto northern hemisphere markets in their winter, as green-ripe fruit are severely damaged at temperatures below 10°C (Meulen 1971). India is finding a high priced market for fresh fruit in the Middle East and this is a possibility for the Australian produce.

As a processed product the mango should find ready acceptance in the world markets.

Cultural details

Propagation

The mango can be multiplied with ease by seed or by several vegetative methods of propagation.

Polyembryonic types come true to type from seed and may be successfully propagated in this manner. Monoembryonic types, however, require vegetative propagation to retain all the desired characters and are generally worked onto rootstocks.

The selection of a suitable rootstock is as important as the selection of scion varieties. It has a strong influence on the growth, yield, and fruit maturity and may be selected for adaptability to soil type and salinity tolerance. Comparatively recently orchards in South Africa and India have been established on nucellar seedlings from polyembryonic mangoes. The Kensington variety in Queensland is polyembryonic and has traditionally been propagated from seed.

There are many ways of grafting and budding mangoes reported in the literature. However, experience in south-east Queensland suggests that actively growing rootstocks, hardened scion material and warm temperatures (25 to 30°C) are a successful combination.

Management

The mango is grown as an orchard tree and can be clean cultivated or grown under sward.

Spacing. This is largely determined by the environment and the vigour of the cultivar. Meulen (1971) recommends 12 m squares under South African conditions with interplants that can be removed after 10 years, while in Queensland spacings of 9 to 15 m x 9 to 15 m are commonly used.

Cropping manipulation. Many techniques have been used to improved productivity. Smudging is an old technique reported from the Philippines for inducing off-season flowering but this has largely given way to chemical induction. There are many reports from India, South Africa and the Philippines on flower induction with ethrel. However, results have been inconsistent in some places and the technique has not been generally adopted at a commercial level. The application of KNO_3 is claimed to be a more reliable inductant in the Philippines where it has been commercially accepted (Bergonia *et al.* 1975).

Irrigation

Successful mango production is generally associated with areas having a pronounced dry season during the winter and spring months with a summer rainfall. Under such conditions irrigation is beneficial from flowering until the 'wet' period begins. To induce heavy natural flowering, irrigation should be withheld for 3 months prior to flowering.

Fertilization

A general criterion with mango nutrition is that care must be taken not to over-fertilize thereby promoting vigour at the expense of fruitfulness. This is particularly true for nitrogen applications.

Meulen (1971) in South Africa and Chadha (1975) in India both agree that timing is critical with bearing trees which should be fertilized at flowering and immediately after harvest. The reasoning here is to supply adequate nutrition for fruit development with the first application and to promote the flower bearing summer flush with the second application. In Queensland, applications of fertilizer are recommended at harvest and again 2 months later.

Various schedules are available but appear to be based on whims rather than experimental results. Ultimately critical levels should be defined and nutrition predicted from leaf analysis.

Pests and diseases

The mango is attacked by several insects, the most important being fruit fly, mango weevil, scale, and gall wasp.

The universal disease of mango is anthracnose *(Colletotrichum gloeosporioides).* The only other disease of any consequence is bacterial black spot caused by an *Erwinia* species.

Harvesting methods

Fruit is gathered by hand from the trees or with the use of picking aids when at a green-ripe condition. Trees are large and mechanical aids such as cherry pickers could be used to assist in harvesting.

Because fruit may be easily bruised bulk handling containers would need careful design.

Marketing characteristics

The fruit matures in the early summer when fruit prices are generally high. The pre-Christmas trade is important. The fruit is attractive in appearance and eagerly sought by those familiar with the fruit. It is felt that a basic problem with the Australian industry is the short production season which reaches 'glut' proportions. Many people in the southern States are unfamiliar with this fruit which would gain greater acceptance with increased familiarity through longer market exposure.

Storage characteristics

The mango has a relatively short shelf life at ambient temperatures. Picked at the mature-green stage, fruit is fully ripe in 7 to 14 days.

The cold storage requirements of mangoes are highly variable ranging from 7.2 to 12.8°C. These temperatures are determined by the cultivar as well as the stage of maturity, production area and cultural practices. However, in general, 12.8°C is considered to be the lowest safe level for commercial purposes.

Fully ripe mangoes for processing may be stored at 0 to 1°C for up to 6 weeks. The skin blackens but the flesh remains in good condition (Anon. 1975).

Apelbaum *et al.* (1977) investigated subatmospheric pressure storage (L.P.S.) of mango and found that prolonged storage life is inversely related to the pressure. Storage was at 13°C and extended storage life was achieved at pressures below 100 mm Hg. At 50 mm Hg fruit damage occurred and no effect was recorded above 250 mm Hg. Spalding and Reeder (1977) found similar results though there were differential varietal responses.

Ethylene may be used to shorten the time from green mature to fully ripe. Fuchs *et al.* (1975) demonstrated that a 100 ppm ethylene treatment for 48 hours at 25°C and 90% relative humidity reduced time of ripening from 10 to 15 days down to 8 to 10 days. It also enhanced fruit colour.

Transport characteristics

The fruit bruises easily and must be handled carefully. In most organized marketing systems fruit is packed in shallow single-layered trays. Removal of field heat to 13°C is desirable to prolong shelf life (Meulen 1971) and in India waxing has also been beneficial in prolonging shelf life (Chadha 1975).

Breakdown of fruit due to anthracnose is a problem with fruit in transit in Queensland and various fungicidal treatments have been investigated. Muirhead (personal communication) recommends a post-harvest benomyl dip 500 ppm active ingredient at 48°C for 5 minutes.

Processing characteristics

The mango can be made into a range of processed products. It can be canned, frozen, and dehydrated as well as used in jams, jellies, chutneys, pickles, juice, nectar and a range of ice confectionery. Processed mangoes in India in 1971 amounted to 16 000 tonnes (Chadha 1975).

Second grade fruit and inferior varieties are processed into varying products in Queensland.

References

Anon. (1975), Horticulture Branch leaflet, *Mango growing in the dry tropics*. Queensland Department of Primary Industries, Brisbane.

Apelbaum, A., Zauberman, G. and Fuchs, Y. (1977), Subatmospheric pressure storage of mango fruits. *Scientia Horticulturae* **7,** 153.

Beal, P.R. (1981), Screening of mango varieties at Bowen, Queensland. *Queensland Journal of Agricultural and Animal Sciences* **38,** 71.

Beal, P.R. and Newman, G.A. (1978), Climatic influences on mango production in the Bowen district in the dry tropics of Queensland. XXth International Horticultural Congress No. 1595.

Bergonia, H.T., Bendad, N.D., Bugante, R.D., Coronel, R.E., Dilay, C.C., Manuel, F.C. and Pantastico, E.B. (1975), 'The Philippines mango'. Philippines Council for Agricultural Research.

Chadha, K.L. (1975), Mango industry in India. Ad hoc government consultation on the improvement and development of the mango industry in Asia and the Far East.

Chaudhri, S.A. (1976), *Mangifera indica* – mango, *in* 'The propagation of tropical fruit trees'. Garner, R.J. and Chaudhri, S.A. (eds), P.D.A. and Commonwealth Agri. Bur., p. 403

Fuchs, Y., Zauberman, G., Yanko, U. and Hemsky, S. (1975), Ripening of mango fruits with ethylene, *Tropical Science* **17,** 211.

Kadman, A., Gazit, S., Ziv, G. (1978), Experiments with the selection of mango rootstocks for the southern Arava. *Israel Journal of Botany* **27,** 34.

Kadman, A. and Oppenheimer, Ch. (1978), Some problems of cultivation of selected subtropical and tropical fruit crops under marginal conditions in Israel. XXth International Horticultural Congress No. 1564.

Lakshminarayana, S., and Aguilar, P.H. (1975), Effect of orchard heating in reducing parthenocarpic fruits in 'Haden' mango. *Proceedings of the Florida State Horticultural Society* **88,** 502.

Malo, S.E. (1977), Mango in Florida. *HortScience* **12,** 286.

Meulen, A. (1971), Mango growing in South Africa. Citrus and Sub-tropical Fruit Res. Inst., Nelspruit Leaflet No. 48, Sub-tropical Fruit Series No. 7, Mango Series No. 1.

Mukherjee, S.K. (1971), Origin of mango *(Mangifera indica). Economic Botany* **26,** 260.

Popenoe, W. (1920), The mango *in* 'Manual of tropical and sub-tropical fruits', a facsimile of the 1920 edition, published by Hafner Press, New York, 1974.

Purseglove, J.W. (1968), Mango, *in* 'Tropical crops. Dicotyledons. 1'. Longmans, London, p. 24.

Rao, M.M., Ravishanker, H., and Bojappa, K.M. (1978), Chemical induction of flowering in fruited shoots of mango *(Mangifera indica* L.*)* trees. XXth International Horticultural Congress No. 1576.

Spalding, D.H., and Reeder, W.F. (1977), Low pressure (hypobaric) storage of mangoes. *Journal of the American Society of Horticultural Science* **102,** 367.

Srivastava, R.P., Chadha, K.L. and Singh, N.P. (1978), Screening of mango rootstocks with special reference to stomatal count and growth characteristics. XXth International Congress No. 1646.

Valmayor, R.V. (1968), 'The mango. Its botany and production'. University of Philippines.

AMBARELLA, HOG PLUM, IMBU, RED MOMBIN

by A.W. WHILEY

Spondias spp.

The genus *Spondias* includes 8 to 12 species indigenous to the tropical regions of the world. Some are used as timber trees. However, only a few have horticultural interest.

Botany

AMBARELLA

Spondias cytherea Sonn.

S. dulcis G. Forst.

The common names attributed to this species are ambarella, wi-tree and otaheite apple. Considered indigenous in Polynesia, the ambarella is an erect, stately, semi-deciduous tree reaching a maximum height of 18 m. The leaves are large, 20 to 30 cm long with subserrate leaflets. Small whitish flowers are produced in large loose panicles 20 to 30 cm long. The fruit is oval, 5 to 7 cm long, orange-yellow, thick skinned, firm fleshed, very juicy with subacid flavour. The seed is large with the flesh clinging tenaciously.

HOG PLUM

Spondias mombin L.

S. lutea L.

Known as the hog plum or yellow mombin, this species is prolific in tropical America. A tall stately tree reaching 18 m in height with large leaves 20 to 30 cm long, with 7 to 17 ovate-lanceolate leaflets. The yellowish white flowers are borne in loose panicles 15 to 30 cm long. The fruit is ovoid, about 2 to 5 cm long, bright yellow with thin skin and an oblong seed of relatively large size. The flesh is yellow, very soft and juicy and of subacid, rather pungent flavour.

IMBU

Spondias tuberosa Arruda

Known as the imbu it is indigenous to the dry plains of north-eastern Brazil. It is a low spreading tree, up to 18 m in diameter, the leaves are 10 to 15 cm long with 5 to 9 oblong-ovate leaflets. The small white flowers are borne in panicles 10 to 15 cm long. The fruit can be likened to a plum, being oval, about 4 cm long and greenish yellow. The skin is thick and tough and the soft, melting, almost liquid flesh suggestive of a sweet orange.

RED MOMBIN

Spondias purpurea L.

S. mombin auth.

The common names attributed to this species are red mombin, Spanish plum, purple mombin. The red mombin is native to tropical America. It is a small, semi-deciduous, spreading tree reaching a maximum height of about 8 m. The leaves are 10 to 20 cm long with subserrate leaflets. The purplish maroon flowers are produced in small unbranched racemes. The fruits, borne singly or in clusters of two or three, are variable in size and form. Commonly they are oval but may be oblong or pyriform, 2.5 to 5 cm long, and yellow to deep red.

Climatic requirements

While indigenous to tropical areas, *Spondias* species will grow and fruit in warmer subtropical climates. While frost tender (slightly more so than the mango), they can all be successfully grown in Florida south of latitude 27°N. Whilst most species are found naturally in dry regions (less than 600 mm rainfall) there is evidence to suggest that they can adapt to wetter, humid environments. In their native environments, flowering occurs during the dry season and this may be important for fruit set.

Soil requirements

Popenoe (1920) mentions briefly the soil requirements of the various species and these are listed below:

- Ambarella: prefers deep rich soils though growth and performance are satisfactory on shallow sandy land.
- Red mombin: the character of the soil is not important as it is found on shallow sands through to heavy clay loams.
- Imbu: little is known regarding adaptability to various soils and while trees are found naturally on very dry gravelly clay loams it should adapt to other soils providing they are well drained

Genetic diversity

With all species there is no known record of named varieties. Superior types exist among diverse seedling populations and selections of some of the better species have been made.

Flowers are bisexual and there are no apparent pollination problems. Hybridization between species is not recorded.

Production season and time of cropping

Little is recorded on this subject but it appears that the trees are early spring flowering with fruit maturing in autumn through to midwinter.

Yield potential

With all species, trees carry heavy crops.

Market acceptability

Various writers confirm that the fruit has an attractive external appearance though comments on the eating quality vary. The ambarella has been compared favourably with some better known tropical fruits by some while others have likened it to a 'very bad mango'. Popenoe (1920) describes the imbu as being the most acceptable of the *Spondias.*

The fruit is consumed fresh or the juice is extracted to flavour drinks. It can also be processed into jellies, jams and sauces.

Export potential

The fruit is unknown on the lucrative international markets. However, reported fruit quality indicates that to date there is little chance of acceptance.

Cultural details

Propagation

Most *Spondias* are easily germinated from seed and improved selections can be vegetatively propagated. They root easily from cuttings of mature wood (large branches may be used for quick establishment).

The species can also be grafted or shield budded onto rootstocks. *S. cytherea* is compatible with *S. mangifera.*

Management
No information is available on management but the species should be able to grow as a standard orchard crop under a grass sward. Spacing would depend on the ultimate size produced in any specific environment.

Irrigation and fertilization
There is a likelihood of response to irrigation during summer and autumn periods; no information is available on fertilization.

Pests and diseases
There are no reports on anthracnose which attacks other Anacardiaceae. However, Sommeijer (1975) reports susceptibility of *S. purpurea* and *S. mombin* to the West Indian fruit fly.

Harvesting methods

In their wild state, trees are climbed or fruit is picked off the ground after it has fallen. Some species growing to 18 m would cause problems if allowed to gain this height in a cultivated situation.

Marketing characteristic (see also market acceptability)

The fruit is marketed at village level as a fresh item.

Storage characteristics

Storage characteristics are unknown but as a fresh soft fruit approximately 5 to 7 days would be normal.

Transport characteristics

The fruit is likely to be sensitive to bruising.

Processing characteristics (see also market acceptability)

Ripe fruit can be used to make jellies, sauces, and jams while unripe fruit can be made into pickles or relishes.

References
Popenoe, W. (1920). Relatives of the mango. *in* 'Manual of tropical and sub-tropical fruits', a facsimile of the 1920 edition published by Hafner Press, New York, 1974.
Purseglove, J.W. (1968). Anacardiaceae. *in* 'Tropical crops. Dicotyledons. I'. Longmans, London, p. 18.
Sommeijer, M.J. (1975). Investigations on the occurrence of the West Indian fruit fly in the Lesser Antilles and Trinidad. FAO Plant Protection Bulletin **23** (6), 185.
Further reading
Morton, J.F. (1961). Why not use and impove the fruitful ambarella. *Horticultural Advance* **5**, 13.

Chapter 4

ANNONACEAE

CUSTARD APPLE

by A.P. GEORGE

Annona spp.
Rollinia spp.

Botany

The family Annonaceae comprises 50 genera. Two genera are of commercial importance, *Annona* comprising 100 species and *Rollinia* comprising 50 species. Both genera contain a large number of species producing edible fruit. The fruit is formed by the fusion of the numerous carpels with the fleshy receptacle and is technically referred to as a syncarpium. Most species of both these genera are semi-deciduous in growth habit and exhibit protogynous dichogamous flowering behaviour.

The main commercial species of *Annona* grown in Queensland and northern New South Wales is *A. atemoya*. Evidence produced by Wester (1915) and Gandhi and Gopalkrishna (1957) suggests that this species originated through the hybridization of *A. cherimola* and *A. squamosa*. Other lesser known varieties of custard apples grown in Queensland may have originated through further interspecific crossing of *A. atemoya* with *A. reticulata*, *A. diversifolia* or *A. muricata*.

While the main commercial varieties of custard apple grown in Queensland and northern New South Wales belong to the species *A. atemoya (cherimola* x *squamosa)*, two other closely related species are grown commercially in other parts of the world, *A. cherimola* in the cooler subtropical regions of California, Spain, Chile, Argentina and Israel, and *A. squamosa* in the tropical regions of India, Malaysia, Thailand and the West Indies. Marked differences in growth and fruiting habits exist between these species. Some of these differences will be discussed in other sections of this paper. The more important species of *Annona* and *Rollinia*, with their potential uses are presented in table 1.

Climatic requirements

Most species of *Annona* and *Rollinia* are strictly tropical in their growth and fruit maturation requirements. These species originated in the lowland regions of northern South America, Mexico and the West Indies. However, a distinctly subtropical species is the cherimoyer *(A. cherimola)*. This species originated in the Andean valleys of Peru and Ecuador at elevations of 2000 m (Popenoe 1920). Varieties of *A. atemoya* which were the result of hybridization of a subtropical and tropical species, appear to have an intermediate climatic requirement.

Temperature

Temperature is the most limiting factor to geographical distribution of the *Annonas* and *Rollinias* with young trees killed by temperatures 3°C below zero. Also of primary importance is the effect of temperatures during the fruit maturation period, with low temperatures retarding fruit maturation and too high temperatures causing fruit to ripen prematurely on the tree and fall. Selection of the correct species and variety for a particular environment is therefore important. The approximate temperature ranges required for optimum growth and fruit maturation of the commercially important species are presented in table 2.

Table 1. Important species of the genera *Annona* and *Rollinia*

Genus *Annona*	Common names	Chromosome number 2n	Gene centre	Potential Uses
A. atemoya Hort.	Atemoya Cherimorinones Custard apple	16 (a)	Various	1. Fresh fruit 2. Rootstock 3. Pollinator species
A. cherimola Mill.	Cherimuyu Cherimoyer Cherimoya Cherimola Custard apple	16 (a) 14 (b)	Andean Valleys of Peru and Ecuador	1. Fresh fruit 2. Rootstock 3.Pollinator species
A. squamosa L.	Sugar apple Sweetsop Sitaphal Ata Custard apple	16 (a) 14 (b)	West Indies	1. Fresh fruit 2. Rootstock 3. Pollinator species
A. reticulata L.	Bullock's heart Ramphala Custard apple (often considered the true custard apple)	16 (a) 14 (b)	West Indies, Mexico	1. Fresh fruit 2. Rootstock
A. muricata L.	Soursop Guanabana Corossol Graviola	16 (a)	West Indies, Mexico	1. Fresh fruit 2. Drinks & purees 3. Rootstock
A. glabra L.	Pond apple Alligator apple	28 (a)	Florida (USA)	1. Rootstock
A. senegalensis Pers.	Wild transvaal apple		South Africa	1. Pollinator species 2. Rootstock
A. diversifolia Saff.	Ilama, Cherimoyer of the lowlands		Mexico, Guatemala, El Salvador	1. Fresh fruit
A. montana Macfady.	Mountain soursop	16 (a)	Cuba, West Indies	1. Fresh fruit 2. Rootstock
A. purpurea Moc. & Sesse	Soncoya, Manirote		Mexico	1. Fresh fruit 2. Rootstock
A. longiflora Aubl.	Wild cherimoya		Mexico	1. Rootstock
A. scleroderma Saff.	Posh te		Mexico, Guatemala	1. Fresh fruit
A. diocia St. Hil.			Paraguay	1. Rootstock
A. globiflora Schlecht			Mexico	1. Rootstock
A. nutans R.E. Fries			Brazil, Paraguay	1. Rootstock

Table 1 (continued).

Genus *Annona*	Common names	Chromosome number 2n	Gene centre	Potential uses
A. fairchildiana Fairchild			Miami, Florida	1. Rootstock
A. testudinea Saff.			Mexico	1. Rootstock
Genus *Rollinia* *R. deliciosa* Saff.	Biriba	48 (a)	Brazil	1. Fresh fruit 2. Rootstock
R. emarginata Schlecht			Brazil, Paraguay	1. Fresh fruit 2. Rootstock

(a) Bowden (1948).
(b) Thakur and Singh (1969)

Table 2. Optimum temperature range for growth and fruit maturation

	Temperature range for optimum growth (°C)		Temperature range for optimum fruit maturation (°C)
Species	Mean max.	Mean min.	Mean max.
A. atemoya	22–28	10–20	20–26
A. cherimola	15–25	7–18	18–22
A. squamosa	25–32	15–25	25–30

Humidity

Ineffective pollination is the most limiting factor to fruit set in the *Annonas*. Hand pollination studies by Schroeder (1943) indicated that fruit set, due to hand pollination, was markedly improved under conditions of moderate temperatures (27°C) and high humidities (80% R.H.) than at higher temperatures (31°C) and lower humidities (30% R.H.). Thakur and Singh (1965) demonstrated that pollen sterility and desiccation occur under these latter conditions. Additional evidence supporting the hypothesis that high relative humidities favour effective pollination is provided by Saavedra (1977) and Ahmed (1936) who demonstrated that the insertion of a drop of water inside the flower at the time of pollination markedly increased fruit set.

Rainfall

Irrigation is essential in most areas particularly during the flowering, fruit set and early fruit maturation periods. Water stress during any of these periods can result in flower or fruit drop. An extended water stress period during the early or late fruit maturation period may result in premature ripening of the fruit while still on the tree.

Wind

The *Annonas* are very susceptible to wind damage particularly during the first 3 years of tree development. Tree shaking, with resultant trunk damage, also appears to be at least partially responsible for the entry and develoment of collar rot organisms. Hot, dry winds during the flowering period cause pollen and stigma desiccation with a resultant reduction in fruit set.

Soil requirements

Although the *Annona* species are capable of growing on a wide range of soil types, from sands to clay loams, higher yields appear to be obtained from trees grown on sand or sandy loam soil types. The reasons for this appear to be due to the indirect influence of poorer soil types on reducing tree vigour, with moderately vigorous trees producing more fruiting laterals and reproductive growth. Good soil drainage is essential with the present range of rootstock species being used. *Annona* species, suitable for use as rootstocks in heavy, poorly drained soils, are presently being evaluated.

Genetic diversity

Extensive genetic diversity exists within all species of *Annona* and *Rollinia.* Existing commercial varieties of *A. cherimola, A. atemoya* and *A. squamosa* exhibit marked variation in growth, fruit setting behaviour fruit size and quality. However, no one variety incorporates all desirable features satisfying productivity and market acceptability requirements. The potential for selecting or breeding better quality, higher yielding varieties is good. Within Australia, improved selections of *Annonas* will most probably be made from within two species, *A. cherimola* which is suitable for cooler subtropical regions, and *A. atemoya* which is suitable for a wide range of subtropical and tropical environments. Rootstock influence on scion behaviour is also quite marked, with genetic variability within seedling rootstock lines and between different rootstock species inducing wide variability in scion performance. Selection of highly productive stock-scion combinations with clonal propagation of both scion and stock is therefore desirable.

Table 3. Some custard apple cultivars

Variety	Tree diam. (m) (mature)*	Time to bearing (years)	Regularity of bearing	Mature tree yield per year (kg)	Fruit shape	Fruit weight (g)	Seed No.** per fruit
Pink's Mammoth	11	6	irregular	120	heart	550	24
Bullock's Heart	11	5	irregular	120	heart	550	24
Island Gem	10	4	irregular	120	round	250	31
Q.A.S.	9	3	regular	150	round	300	31
Nielsen	9	3	regular	150	oval	300	31
African Pride	8	2	regular	180	heart	300	36

* Varieties grafted on *A. cherimola* rootstock.

** Seed no. per 500 g fruit.

Production season and time of cropping

The main harvesting period for *A. atemoya* in south-eastern Queensland is between April and early August with the peak harvest period varying with variety and environmental conditions during the flowering and fruit setting period. In northern New South Wales, due to cooler growing conditions, fruit tends to mature up to 1 month later than similar varieties growing in south-eastern Queensland. Management practices such as defoliation and summer pruning may also alter the harvesting period. In terms of precosity of bearing, marked differences exist between varieties and variety-rootstock combinations. Management factors can also alter this appreciably.

Yield potential

Erratic and poor fruit set have been major problems limiting *Annona* production and further expansion in many parts of the world. Unfortunately, in Queensland, poor fruit set is a problem characteristic of our better quality varieties. The causes of poor fruit set are diverse but fall into the two major categories: ineffective pollination, and lack of reproductive growth.

Ineffective pollination has been found to be due to a high degree of pollen sterility and low pollen germination. Environmental factors also markedly affect the degree of overlap of male and female stages (protogynous dichogamy) required for self-pollination and fruit set.

Lack of reproductive growth is a complex problem associated with juvenility, excessive tree vigour, apical dominance and bud dormancy effects. Genotype, environment and management factors including nutrition, pruning and defoliation, all influence this problem. A discussion of these factors and their influence on yield is outside the scope of this paper.

Under optimum conditions, 30 tonnes per hectare of marketable fruit *(A. atemoya)* have been obtained. This is equivalent to 3500 half bushel cases per hectare obtained from an orchard planted at a density of 150 trees per hectare.

Market acceptability

Market acceptability is high with well-presented fruit.

Export potential

Export potential is difficult to assess at this stage. However, due to the relatively high perishability of the fresh fruit, efficient transport and storage techniques would need to be developed.

Cultural details

Propagation

Due to the high degree of variability of seedling *Annona* lines, clonal propagation of both scions and rootstocks is highly desirable. The *Annonas* are easily budded or grafted using the whip, whip and tongue, or side cleft graft. Clonal propagation by cuttings of some *A. atemoya* varieties is relatively easy. Time of propagation with respect to percentage strike is highly critical (table 4).

Management

Management techniques to ensure high yields in the *Annonas* are presently being developed. The establishment of wind-breaks prior to and during the early tree development phase is essential if structural tree damage is to be prevented. Pruning and defoliation encourages the development of small fruit bearing laterals. Pruning is generally carried out when trees are dormant (early September) and in summer (October and December). Defoliation using 500 ppm ethephon or 25% urea solutions, appears to be most effective if applied just prior to natural budbreak.

Irrigation

Irrigation is essential to prevent water stress during the flowering and fruit set period. Some humidity modification at the peak flowering period may be feasible with some irrigation systems, with maintenance of humidities between 70 to 80% being desirable to ensure effective pollination.

Table 4. Clonal propagation by cuttings

A. atemoya var. Pinks Mammoth

Treatment	Time of removal							
	April		June		August		October	
	% call.**	% rooted*	% call.	% rooted	% call.	% rooted	% call.	% rooted
Control (tip cuttings)	75	25	80	30	95	35	45	20
I.B.A. 1000 ppm	65	5	90	50	95	60	40	20
I.B.A. 2000 ppm	65	0	90	35	95	80	20	5

A. atemoya var. African Pride

Treatment	Time of removal							
	April		June		August		October	
	% Call.**	% rooted*	% call.	% rooted	% call.	% rooted	% call.	% rooted
Control (tip cuttings)	75	30	80	45	95	35	70	35
I.B.A. 1000 ppm	80	35	95	60	80	50	70	60
I.B.A. 2000 ppm	85	35	75	50	90	60	75	50

* All cuttings propagated in sand, mist beds held at a constant bottom temperature of 26°C.
** call. = calloused.

Fertilization

Fertilizer rates and timing need to be adjusted according to tree age, variety and soil type. With mature trees, excessively vigorous trees produce fewer fruiting laterals and reduced reproductive growth. Adjustment of fertilizer rates, in particular nitrogen, to obtain a moderate level of tree vigour is required. Applications of nitrogen just prior to and after the commencement of flowering appear to enhance fruit set.

Pests and diseases

Pests and diseases are generally not major problems of *Annonas* grown in Australia. The most serious pest, which is more prevalent in some years and districts, is the citrus mealybug *(Planococcus citri)*. Biological control measures for this pest are presently being evaluated.

The most serious disease problem is a collar rot disorder affecting mainly young trees. The cause of this disorder is not known. It appears to be more prevalent after prolonged windy periods, when some damage to trunk tissues at ground level may be caused. Of the fruit diseases, black canker *(Phomopsis anonacearum)*, purple blotch *(Phytophthora palmivora)* and diplodia rot *(Botryodiplodia theobromae)* do not warrant control measures. A description of these diseases is given by Purss (1953).

Harvesting methods

Mature fruit are harvested while still firm on the tree, when the interstices between carpels are cream in colour. Harvested fruit are generally graded and packed into 18 L, polystyrene containers prior to shipment.

Marketing characteristics

Fruit generally ripens 3 to 7 days after harvest, becoming soft and easily damaged. For this reason, rapid and efficient handling and transport procedures are required.

Storage characteristics

The *Annonas* are very susceptible to cool storage damage, with temperatures below 6°C for periods greater than 3 days resulting in surface skin discoloration. Broughton and Guat (1979) defined ideal storage requirements for *A. squamosa,* a climacteric species. These researchers recommended that storage conditions be maintained between 15 and 20°C, under low oxygen and ethylene tensions, a 10% carbon dioxide level, and a relative humidity of 85 to 90% in the storage atmosphere. Shelf life of fruit stored under these conditions is 13 days.

Transport characteristics

While still firm the *Annonas* can be readily transported considerable distances, as demonstrated by the high percentage of Queensland produce marketed in Sydney and Melbourne. Pre-cooling of the fruit, prior to shipment, appears to improve shelf life.

Processing characteristics

With the possible exception of the soursop *(A. muricata),* the *Annonas* do not readily lend themselves to processing, mainly due to the high sugar content and delicate, aromatic flavour of the fresh fruit product. The soursop is manufactured into a wide range of drinks and purees in several central American countries.

References

Ahmed, M.S. (1936), Pollination and selection in *Annona squamosa* and *A. cherimola. Bul. Min. Agr. Egypt Tech. Sci. Serv. Hort. Sect.* **157,** 1.

Bowden, W.M. (1948), Chromosome numbers in the Annonaceae. *American Journal of Botany* **35,** 337.

Broughton, W.J. and Guat, T. (1979), Storage conditions and ripening of the custard apple, *Annona squamosa* L. *Scientia Horticulturae* **10,** 73.

Gandhi, S.R. and Gopalkrishna, N. (1957), The *atemoya,* a promising new *Annona* hybrid. *Proceedings of the Indian Academy of Science* **45,** 234.

Popenoe, W. (1920), The Annonaceous fruits, *in* 'Manual of tropical and sub-tropical fruits', a facsimile of the 1920 edition, published by Hafner Press, New York, 1974.

Purss, G.S. (1953), The fruit rots of the custard apple. *Queensland Journal of Agricultural Science* **10,** 247.

Saavedra, E. (1977), Influence of pollen grain stage at the time of hand pollination as a factor on fruit set of cherimoyer. *HortScience* **12,** 117.

Schroeder, C.A. (1943), Hand pollination studies on the cherimoya. *Proceedings of the American Society for Horticultural Science* **43,** 39.

Thakur, D.R. and Singh, R.N. (1965), Studies on pollen morphology, pollination and fruit set in some *Annonas. Indian Journal of Horticulture* **22,** 10.

Thakur, D.R. and Singh, R.N. (1969), Karyomorphological studies in some *Annona* species. *Indian Journal of Genetics and Plant Breeding* **29,** 285.

Wester, P.J. (1915), Hybridization of *Annonas. Philippine Agricultural Review* **8,** 10.

Chapter 5

APOCYNACEAE

NATAL PLUM, AMATUNGULA, CARANDA, AUSTRALIAN CARISSAE

by D. H. MAGGS

Carissa spp.

The Apocynaceae form a distinctive family with a floral structure that favours a high degree of self-fertilization through the close proximity of the ring of stamens to the stigma. It is of considerable economic importance, including latex species *(Landolphia, Funtumia)*, drug species *(Acokanthera, Rauwolfia, Strophanthus)*, ornamentals *(Allamanda, Plumeria, Nerium)*, and the plum-fruited *Carissa*.

Botany

NATAL PLUM
Carissa grandiflora (E.H. Mey) A. DC.
C. grandiflora is a spiny evergreen shrub forming dense bushes, prostrate or up to 5 m high. Leaves are entire, round, sometimes pointed, thick, leathery, bright green, purple in cold winters. The spines between the leaf pairs are doubly branched, stout, firm and sharp. Flowers are in terminal cymes, 3 to 5 cm in diameter, white, strongly scented, perfect. Fruits are bright red, 3 to 5 cm long, ovoid to spherical; skin thin; pulp with milky juice; seeds flattened, few to many; mainly ripening late summer, but throughout the year in humid climates such as in Hawaii. The natal plum is a South African species, where it is marketed as a summer fruit. It is widely distributed throughout the tropics as an ornamental and effective hedge plant.

AMATUNGULA
Carissa bispinosa (L.) Desf. ex Brenan
Carissa arduina Lam; *C. acuminata* DC.
C. bispinosa is much like *C. grandiflora.* The flowers are smaller, 2 to 3 cm in diameter; fruits are ovoid, 1.5 cm long with about 6 seeds. Origin of the species is southern central Africa.

CARANDA
Carissa carandas L.
This is a shrub or small tree, sometimes straggling. Leaves are oval, 4 to 8 cm long. Flowers are 2 cm in diameter. Fruit are ovoid, about 2 cm long, at first red, ripening black to purple. Originating in India, it is now widely distributed in the humid tropics.

AUSTRALIAN CARISSAE
Carissa lanceolata R. Br.
This is an erect, wiry shrub to 1.5 m high. Branches are slender, stiff, 1 to 3 mm in diameter, prostrate branches rooting at a node. Leaves are 2 to 3 cm long x 3 mm wide, yellow-green. Spines are inconspicuous, usually in pairs, slender, up to 1.5 cm long. Flowers are 7 mm in diameter, crowded in the leaf axils, often on short branches. Fruits are black, 5 mm in diameter. Origin of the species is central and northern Australia.

Carissa lowii
Similar to *C. lanceolata.* Origin, central Australia (P. Latz).

Carissa ovata R. Br.
Carissa brownii F. Muell.
C. ovata is an erect, much branched shrub to 1.5 m high. Leaves are round to rhomboidal, hard. Flowers are 3 to 5 mm in diameter. Fruit are ovoid, 10 to 15 mm long, black. The species is distributed along the east coast of Australia, northern New South Wales, and Queensland.

Climatic requirements

Carissae grow in dry, subtropical regions. *C. grandiflora* is adapted to more extreme conditions and can survive –5°C; *C. carandas* tolerates a wetter, warmer climate.

Soil requirements

Carissae are not exacting in their soil requirements, *C. carandas* preferring drier soils. *C. grandiflora* equals *Bougainvillea* and *Rosemarinum* in salinity tolerance, that is, highly tolerant.

Genetic diversity

C. grandiflora varies greatly in growth habit, cold tolerance, spinal morphology, self-compatability and fruit characters. A number of cultivars were selected in California for fruit characters, but have now been lost (Thompson 1976). Current interest in this species is as an ornamental ground cover for hot, dry situations.

Production season

All species apparently carry their main crops in summer. Off-season fruits are freely produced under favourable growing conditions.

Yield potential

No data are available. As *C. grandiflora* is a commercial crop in South Africa, 3 t ha^{-1} could be taken as a minimum value. In central Australia *C. lowii* can be picked by the handful.

Market acceptability

C. grandiflora is in demand on South African markets.

Cultural details *(C. grandiflora)*

Propagation is by soft and semi-hardwood cuttings, budding, grafting, seedlings. Cuttings sucker badly and below ground buds should be extirpated when taking cuttings. *C. carandas* as a rootstock is said to increase fruitfulness and reduce plant size of *C. grandiflora*.

Carrisae should be pruned to a narrow vertical hedge to increase accessibility of the fruit.

Water of 8 mmho conductivity (c. 5000 ppm) is acceptable in irrigation. (Bernstein *et al.* 1972).

Harvesting methods

Hand picking has problems because the doubly branched thorns usually force the operator to tear the fruit from the pedicel instead of below it. Australian species could be suited to mechanical harvesting.

Marketing characteristics

Carrisae are marketed mainly as fresh fruit. Culls are used for processing. Dried latex may mar the appearance of fresh fruit.

Processing characteristics

Fruit are used in jams and sauces similar to cranberry sauce. Fruit composition of *C. grandiflora* is water 78%, protein 0.6%, fat 1.0%, sugars 12%, acid 1.2% and fibre 0.9%. As with figs, latex disappears with cooking.

References

Bernstein, L., Francois, L.E., and Clark, R.A. (1972), Salt tolerance of ornamental shrubs and ground covers. *Journal of American Society of Horticultural Science* **97**, 550.

Thompson, P.H. (1976), The carissa in California. *California Rare Fruit Growers Yearbook* **8**, 73.

Further reading

Popenoe, W. (1920), 'Manual of tropical and sub-tropical fruits', a facsimile of the 1920 edition, published by Hafner Press, New York, 1974.

Sturrocks, D. (1959), 'Fruits for southern Florida'. Stuart, Florida.

Uphof, J.C.T. (1968). 'Dictionary of economic plants'. 2nd edition, Cramer, Lehre.

Chapter 6

BOMBACACEAE

DURIAN
Durio zibethinus J. Murr.

by B.J. WATSON

Botany

In the family Bombacaceae, the genus *Durio* consists of 27 species at least 6 of which produce an edible fruit. Natural distribution is peninsular Malaysia, southern Thailand, Indonesia and Borneo, the latter having the greatest range of species (Soegeng-Reksodihardjo 1962). The most widely cultivated species is *Durio zibethinus* J. Murr., commonly referred to as durian or in some odd cases, civet fruit, the fruit maligning the civet cat's odour or vice versa.

Durian in prime environment grows some 20 to 45 m tall but grafted clones seldom exceed 20 m. Cultivars usually assume a Christmas tree habit but seedlings tend toward a tall branchless trunk with an irregular and dense crown. Leaves are simple, drooping and alternate, 80 to 200 mm long, light green on the upper surface and rusty brown on the lower. Flowers arise in stalked pendulous bunches with up to 25 individual flowers some 50 to 70 mm long on the main and smaller lateral branches. The fleshy epicalyx splits to reveal 5 united sepals and 5 yellowish white petals. Durian flowers open from 3 p.m. to late evening. Before midnight most pollen is shed and all flower parts excepting the pistil fall to the ground. Moths and nectarivorous bats are probably the most important pollinating agents.

The fruit (usually only 1 or 2 developing from each flower cluster) is stalked, pendulous, round to oblong, commonly 200 mm long by 175 mm thick, but may be up to 400 mm long for some Thai cultivars. The rind is tough and covered with thick sharp pointed coarse spines. The fruit is actually a capsule and splits when ripe into 3 to 4 segments of irregular width, each of which contains from 1 to 7 seeds. Each seed is embedded in a rich cream or yellow coloured pulp (the aril). The pulp is the edible portion and varies in colour, texture and thickness between seedlings or cultivars. Seeds are large, 40 to 70 mm long by about 25 mm thick. The rind of the fruit is extremely tough and has evolved so as to withstand impact without major damage from a fall of as much as 40 m. The fruit is attractive to a great variety of animals, including amongst many, the orang-outang, elephant, tiger and even the domestic cat.

Climatic requirements

In latitudes north of the equator durian is cultivated at up to 18° in Burma, Thailand, Vietnam and India but is more popularly and successfully grown within 15°. Compared with most other 'truly tropical fruits' durian is surprisingly accommodating in tolerance to 'low' mean temperatures. Terra (1952) lists five centres in Java between 400 and 600 m altitude where durian is grown, and one special centre for durian production is Puspo in East Java, at an altitude of 650 m. However, although relatively cool at this altitude (mean yearly normal temperature of 23°C as compared with 24.5°C at Cairns), there are no great extremes as there are at say 15 to 17° N or S of the equator, particularly in areas with continental weather influences. At Kamerunga Research Station trees are actively vegetative in all months except June, July and August, which suggests that growth is limited at mean monthly temperatures of below 22°C. Occasional temperatures of below 10°C appear to cause premature leaf abscission. At the other extreme, trees appear to tolerate temperatures of up to 46°C in areas of Thailand and India.

In most Asian production areas mean yearly rainfall exceeds 2000 mm and historically the better production sites have developed in areas of rainfall well distributed throughout the year and with mean yearly totals exceeding 3000 mm.

Malo and Martin (1979) state that established durians cannot withstand more than 3 months drought without suffering irreversible damage.

There is no doubt however, that with irrigation drier zones can produce good crops.

Soil requirements

The durian grows best in a rich deep well-drained sandy clay or clay loam. Heavy clay soils with poor drainage should be avoided since the root system is very sensitive to standing water, and such conditions are conducive to proliferation of the most devastating durian disease, *Phytophthora palmivora*.

Prabhadhapongse (personal communication 1979) a leading grower in Khlung, Thailand, states that organic animal manures should be avoided since they are conducive to *P. palmivora* infestation. In nursery conditions in north Queensland, seedlings sustain markedly more vigorous growth when potted into light sandy rather than high clay content mediums.

Genetic diversity

Durian progeny are very heterozygous in all attributes. Evidence provided by Valmayor *et al.* (1965) and Malo and Martin (1979) confirms previous opinions that most durian seedling trees have a high degree of self incompatability and flowers must be cross-pollinated from other trees in order to set fruit. This would add to the genetic diversity. However, it is possible that clonal selection has resulted in a number of cultivars which are self compatible. Prabhadhapongse (personal communication) is adamant that in Thailand, in large plantings of one cultivar, individual trees do not produce less than mixed cultivar plantings. This observation is supported by pollination studies by Soepadmo and Eow (1977) in Malaysia. There are marked differences in: tree form, vigour and leaf size; flowering position, flower shape and fruit number per 'cluster'; peduncle length; fruit size and shape; rind colour and thickness; aril volume, colour, aroma, texture and flavour; seed number and size. Selection criteria are usually in decreasing order of importance of: aril volume, texture and flavour, yield per tree, spread of harvest season and shelf life.

Malaysia, Singapore and Thailand have selection programmes backed by government agencies which have covered hundreds of cultivars. However, short lists of current recommendations seldom exceed 10 cultivars. It is interesting that even though Thailand has very little genetic base in natural distribution it now is regarded as having the best cultivars. Indonesia is relatively inactive in current selection work.

At Kamerunga, three Malaysian, one Indonesian and five Thai cultivars have survived quarantine and field establishment.

Recommendations for best production/fruit quality and seasonal range cultivars in various countries are as follows:

Malaysia — D2 (Dato Nina), D7, D10 (Durian Hijau), D16, D24, D98 (Katoi), D116 (Batu), D117 (Gombak), D123 (Rim 2) and Hew 3.

Thailand — Gaanyao, Montong, Kob Picul, Chanee, Luang, Kradoom Tong, Chompoosri.

Indonesia — Sitokong, Bakul, Mas, Sitebel, Simanalagi, Simadat.

Limited data on cultivar characteristics are available. The table below lists data for a few Malaysian cultivars.

Some Malaysian durian cultivar characteristics

Cultivar	Mean tree height at 8 years (m)	Mean tree diam. at 8 years (m)	Mean accumulated No. of fruit/tree at 8 years	Mean fruit weight (kg)
D2	6.1	3.25	—	1.36
D8	5.8	2.85	211	1.22
D24	7.5	3.05	115	1.23
D66	7.3	2.45	115	1.09
D88	5.5	1.85	115	3.0
D96	6.5	2.35	10	—
DR	7.8	2.85	262	—

Production season and time of cropping

Durian is not photoperiod or temperature responsive for flower initiation in equatorial zones. However, in areas from 10° to 18° north or south of the equator flowering normally commences in the spring months, with harvest in midsummer to autumn. In areas of Malaysia and Sumatra where there are no defined wet and dry seasons, individual trees produce 2 crops a year with peak harvests in June and December respectively.

In tropical areas with marked seasonal wet and dry seasons, for example, India and eastern Java, flowering commences near the end of the dry season. In north Queensland there has been limited flowering in the October-December period. On the basis of a normal development period of 110 to 150 days in Asia one might expect a harvest period between January and May in northern Australia. Fruit growth follows a sigmoidal pattern. Cultivars generally commence bearing in the fourth to sixth year after field planting whereas seedlings take from 7 to 10 years.

Yield potential

Thailand produced 469 000 tonnes of durian fruits in 1975 with an average yield of 8.3 tonnes per hectare (Vangnai personal communication 1979). Thailand is the world's leading producer followed by Indonesia with 157 000 tonnes.

Fruit weight varies from 1.5 to 4.0 kg for commercial grades but odd fruit up to 8 kg are common. Good orchards in Malaysia and Thailand commonly produce 10 to 18 tonnes per hectare per annum in the tenth to fifteenth year from planting at which time initial densities of 100 trees per hectare are reduced to approximately 50 trees. There is a slight tendency for biennial bearing even in areas where there are two crops per year. Yield potential in northern Australia cannot be estimated at this stage.

Market acceptability

To the average Thai or Malaysian a query as to market acceptability of durian would bring a similar reply to that asked of the average European as to the acceptability of the peach or the apple. However, the appearance, smell and flavour of durian is abhorrent to many if not most Europeans at first experience. The most subtle approach to promoting the fruit would be to introduce it as a flavouring of ice cream and sweets. However, how wide an appeal the fresh fruit would have to Australians cannot be gauged although it would probably be initially acceptable (that is, eaten more than once) to less than 10% of the population.

There would, however, be a substantial and demanding market in the Asian born population and returned expatriates from Malaysia and Thailand etc. However, there would be obvious problems in transporting and marketing the fruit due to the 'aroma' which to many people is quite nauseating.

Export potential

It is unlikely that cost of production (if indeed durian were proven to crop well in some areas of northern Australia) would allow an export trade in durian with the traditional European market (mainly in France, for the very large expatriate Vietnamese population) which is currently supplied from Thailand. However, Australian production could be earlier in the year than the main Thai season and very good prices are obtained in Asian markets for out-of-season fruits. Overall, export potential must be regarded as minimal until production is proven.

Cultural details

Propagation

Asexual propagation of durian is not difficult and a large number of methods are used. In Malaysia, Indonesia and the Philippines the modified forkert bud graft, wedge, whip and approach grafts are most common. However, the Thais have developed a type of bottle graft with which approximately 90% of their trees are produced. Durian seedlings are sown in beds and dug up when approximately 300 to 600 mm high. The roots are packed with moist coconut husk fibre and placed in a sealed polythene bag. This seedling is then tied to a branch adjacent to a suitable scion lateral. The top of the stock is decapitated and inarched in the fashion of a side veneer graft. After the union is complete the scion is cut at about the level of the bottom of the seedling roots. This scion butt when potted up develops roots and the grafted tree does in fact produce two root systems. In Thailand seedlings of Chanee variety are reputed to make very good rootstocks. There is increasing use of the *Durio* species *D. malaccensis, D. mansoni* and *D. lowianus* for resistance to *Phytophthora palmivora.* However, there are reports of scion overgrowth with some clones (Vangnai personal communication 1979).

In India a relative of the durian, *Cullenia excelsa* (native of Sri Lanka) is reputed to make a good vigorous rootstock and increases precocity of fruiting.

Management

Following planting in Thailand trees are provided with temporary shade for the first year. This is also necessary in north Queensland. In Thailand and Malaysia trees are left to grow unrestricted for the first 2 or 3 years but thereafter the interior of the tree is thinned out removing all thin or dead branches and water shoots. Cut surfaces are treated with a fungicide and bitumastic compound. The under-tree and drip line area is kept free of weeds, principally to avoid a microclimate suitable for *Phytophthora palmivora* proliferation. For the same reason the trunk of the tree is kept clean (water shoots and weak branches) up to 1 m above the ground.

Planting distances are commonly 8 m (156 trees per hectare) to 16 m (40 trees per hectare) apart but at the former spacing tree thinning is necessary in the eighth year. Tropical legume cover crops between trees are recommended. During fruit development, limbs are often propped to prevent breakage.

Irrigation

In Asia areas with a dry season of greater than 3 months are regarded as marginal for durian but with adequate irrigation this limitation is removed. Supplementary watering for at least 6 months in late winter, spring and early summer is necessary in north Queensland. Periods of most critical water requirement are at flowering and during fruit development. However, the popular variety 'Chanee' in Thailand tends to drop flowers if watered to the same extent as other varieties.

1 Cashew — Apple and Nut

2 Mango — Kensington

3 Mango — Tommy Atkins

4 Custard Apple — Pinks Mammoth

5 Rollinia

6 Custard Apple — African Pride

7 Soursop

8 Durian

9 Matisia

10 Chinese Olive

11 Pili Nut

12 Mabolo

13 Black Persimmon

14 Mafai

15 Rambai

16 Chermai

17 Chermai

18 Mangosteen

19 Mammea

Fertilization

Nutrient removal by the crop is relatively low. Total element removal in Malaysia is shown in the following table.

Nutrient removal (kg/ha) from harvest of 7000 kg of fruit per hectare per annum.

N	P	K	Ca	Mg
16.4	2.9	28.5	2.0	3.4

(Ng and Thamboo 1967)

Malaysian fertilizer recommendations (Kanapathy 1976) for average soils are N:P:K mixtures approximating 14:4:3.5 for the first 5 years and thereafter 12:4:7 twice yearly in the progressive increase range of 100 g to 4 kg per tree per annum from planting up to 12 years of age. Dolomite is also applied. Up to 15 g P is added to the planting hole.

In Thailand (Prabhadhapongse, personal communication 1979) fertilizers are applied in the drip line and chipped in; as flowers emerge N:P:K — 1:1:1; just after harvest 2:1:2; and 4 months later 1:1:1. Additional animal manures are favoured in Malaysia, but not in Thailand.

Pests and diseases

The major disease problem is patch canker caused by *Phytophthora palmivora* which devastates plantings throughout Asia. The organism is a primary parasite of durian roots, followed by canker development on the trunk at or just above ground level which eventually girdles the tree. Control is mainly cultural, good drainage, removal of vegetation around the trunk to prevent a humid microclimate, use of resistant rootstocks and inarching one or more 'nurse' rootstocks. Use of the newer systemic fungicides, Ridomil (R) etc. has been initiated. Other diseases of importance are root rot caused by *Pythium complectens* and pink disease, *Corticium salmonicolour.* In Australia the fungus *Macrophomina phaseoli* caused decline and death of trees of the cultivar D24 at Kamerunga Research Station.

The tree and fruit is relatively free of serious insect pests, but in north Queensland damage by the black beetle *(Rhyparida discopunctata)* is severe on leaves of young trees. Monkeys and apes are a problem with ripe fruit in Malaysia.

Harvesting methods

In Malaysia fruit are allowed to fall on the ground but picked up at least once daily. The Thai method is to detach the fruit (with the peduncle intact) before maturity, but at a stage late enough to ensure normal ripening without loss of flavour. The peduncle is often wrapped with a banana leaf or paper which reputedly extends the shelf life.

Marketing characteristics

Since the pericarp shows no colour change at any stage, maturity is gauged by smell or the sound produced by tapping the spines. Malaysian fruit are normally marketed the morning following collection from the tree and since the fruit locules are often separated on impact with the ground the fruit deteriorates quickly and the aril may be rancid within 36 hours. However, sound unbroken fruit remain edible for 2 to 3 days. Thai fruit harvested from the tree often have a shelf life extending to 1 week and very considerable quantities are exported to Malaysia, Singapore, Hong Kong and Europe.

Packaging for market is simplistic. The hard rind prevents major damage and fruit are usually separated by banana leaves in bamboo baskets, or bulk loaded onto trucks or small boats. For export, large cardboard cartons with capacity up to 50 kg are the norm.

Storage characteristics

Durian is very seldom refrigerated in Asia mainly because it is a bulky fruit and markets are not sophisticated. However, sound fruit will store satisfactorily for up to 3 weeks at 15°C. Bauchau (1972) states that the fresh pulp (aril) can be kept up to 3 months by quick freezing at a temperature of minus 24°C without change in flavour.

Transport characteristics

Because of rapid deterioration, mature fruit are transported to the market site as quickly as possible.

For any carrier to accept fruit in Australia airtight containers would probably be necessary to avoid 'air pollution'. However, some cultivars have relatively 'mild' fruit with very little objectionable odour.

Processing characteristics

A great number of preparations using durian are made in Asia including dodol (durian pulp mixed with coconut milk, glutinous rice and palm sugar), sugary cakes, etc. However, for a European market it would be most acceptable as a flavouring for ice cream, sweets, soft drinks and pastry. The aril can be preserved for flavouring by spray or drum drying, but is very hygroscopic in this state and must be stored in airtight containers.

Durian seeds (high in starch, oil and protein) can be roasted or boiled and have a chestnut flavour which is very acceptable. The aril and seeds of durian represent from 14 to 22% and 18 to 22% of the fruit weight respectively.

The distinctive aroma of durian is attributed to thiols, thioethers and esters (Staton 1966) and the compounds responsible for the objectionable smell are hydrogen sulphide and diethyldisulphide.

The pulp (aril) contains:

Water	58%	Minerals	1.2% including:
Protein	28%	Calcium	0.01%
Fat	3.9%	Phosphorus	0.05%
Carbohydrate (about 12% sugars and 12% starch)	34.1%	Iron	0.001%
		Carotene	20 Int. Vit. A units
		Vitamin C	25 g per 1000 g

References

Bauchau, P.C. (1972), Utilization of local Malaysian fruits: development of some products from durian. F.A.O. Work Paper 69.

Kanapathy, K. (1976), Guide to fertilizer use in Peninsular Malaysia. Bulletin 143, Ministry of Agriculture and rural Development, Malaysia, p. 88.

Malo, S.F. and Martin, F.W. (1979), 'Cultivation of neglected tropical fruits with promise. Part 7. The durian'. USDA Science and Ed. Admin.

Ng and Thamboo (1967), Nutrient removal studies on Malayan fruits: Durian and Rambutan. *Malay Agricultural Journal* **46,** 164.

Soegeng-Reksodihardjo, W. (1962), The species of durio with edible fruits. *Economic Botany* **16,** 270.

Soepadmo, E. and Eow, B.K. (1977), The reproductive biology of *Durio Zibethinus. Gardens Bulletin* (Singapore) **29,** 25.

Stanton, W.R. (1966), The chemical composition of some tropical food plants. VI Durian. *Tropical Science* **8,** 6.

Terra, G.J.A. (1952), Some ecological requirements of Indonesian fruit trees. *Landbouw.* **24,** 193.

Valmayor, R.V., Coronel, R.E. and Ramirez, D.A. (1965), Studies of floral biology, fruit set and development in durian. *Philippine Agriculturist* **48,** 355.

Further reading

Kochummen, K.M. (1972), Bombacaceae *in* 'Tree flora in Malaya'. T.C. Whitmore (ed.), Longman Malaysia Sdn. Bhd., Kuala Lumpur.

MATISIA (syn. SOUTH AMERICAN SAPOTE)

by I.L MILLER
and B.J. WATSON

Matisia cordata Humb. & Bonpl.
Quararibea cordata Humb. & Bonpl.

Botany

Fruits with the common name 'sapote' belong to a wide range of botanical families. The plant discussed in this paper, the Matisia cr South American sapote, is a member of the family Bombacaceae.

Humboldt and Bonpland were the first to describe the species on an expedition in Colombia in 1801 (Romero Castaneda 1961). Hodge (1960) reported that while it is generally placed in the genus *Matisia* it has been assigned synonymy under the genus *Quararibea* and the combination *Quararibea cordata* Humb. & Bonpl. Garcia-Barriga and Hernandez, was made in 1952. However, Uphof (1968) refers to it as *Matisia cordata*, and a botanist at the Caracas Botanical Institue, Venezuela, says it was originally described as *Quararibea cordata* Humb. & Bonpl. but is now known as *Matisia cordata* Humb. & Bonpl. (J. Steyermark personal communication 1980).

It has a number of common names: South American sapote (Hodge 1960); zapote chupachupa (Perez-Arbelaez 1956); chupachupa (Romero Castaneda 1961); sapote colombiano (F. Leal personal communication 1980); and firolisto (M. Figueroa personal communication 1980).

M. cordata is a native of the hot lowland forests of the lower Andes in Colombia, Venezuela, Ecuador and Peru (Perez-Arbelaez 1956; Hodge 1960).

According to Romero Castaneda (1961) it is a large tree growing up to 20 m high, although Hodge (1960) confirms the observation of Popenoe (1924) that under cultivation it reaches an ultimate height of 12 m. Whorls of branches (often 5) originate from the main trunk bearing terminal rosettes of large leaves up to 45 cm long and 40 cm wide, on long petioles.

Flowers emerge from mature branches in groups varying from 3 to 10 per cluster and are 6 to 7 cm long. The calyx is large and green occupying about three-eighths of the total length of the flower. The corolla is slightly larger than the calyx consisting of 5 creamy coloured petals. The fruits are ovoid, up to 12 cm long and 8 cm wide, with a persistent calyx adhering to the fruit. The skin is hard and thick, greeny brown, with a felt-like surface. Below the skin is a yellow fibrous pulp of sweet and agreeable flavour containing 5 seeds with adhered fibre. The flavour is described as a composite carrot, melon, mango, papaw. Fruits mature in 6 to 8 months from flowering.

Climatic requirements

As mentioned above *M. cordata* is found naturally in the rainforests of the lower Andes. It is grown in elevations of up to 1200 m (Perez-Arbelaez 1956). Based on its latitudinal and altitudinal range in South America it would no doubt grow well in the climate of northern Australia provided adequate water is provided in the dry season. It has been reported by Hodge (1960) that it would probably not thrive in subtropical areas such as Florida. However, Whitman (1976) reported first fruiting in Florida in 1973 on 9-year-old trees obtained as seed from Iquitos in the Amazon Basin in Peru. Plants observed out of their natural habitat in drier areas of Venezuela (850 to 1000 mm average rainfall) looked poor despite some dry season irrigation although obviously not enough. A report from the Cairns area (L. Muller personal communication 1980) suggests that it needs plenty of water and heat in the early stages of growth together with shelter from wind. Trees at Kamerunga Horticultural Research Station (ex Amazon Basin) have made 5.5 m of growth in 4 years. Growth proceeds uninterrupted throughout the year but slows down somewhat in June, July and August, in north Queensland.

Soil requirements

There is very little documented information. Soils would probably need to be deep, well drained, with a high organic matter content. Florida grown trees apparently tolerate pH up to 8 but prefer a lower level.

Genetic diversity

Hodge (1960) reports that in its area of cultivation no selection of superior strains had been made up to that date and all trees in culture were seedlings. However, Whitman (1976) suggests that there is a range of fibre content within the gene pool and this is inherited. The Amazon Basin tree is considered a superior strain to that originating in other areas. There is now a programme of cultivar selection at Tefe on the Amazon. (Clement, personal communication 1980).

Production season and time of cropping

It was reported by Hodge (1960) that whether there is a regular flowering season is unknown though it probably depends on local conditions. He observed flowers in late July in northern Peru and in February at Medellin, Colombia, locations which are equidistant from the equator (that is, 6°S and 6°N respectively). At Medellin fruits matured 8 months after flowering (that is, October). Perez-Arbelaez (1956) reported that at Mariquita, Colombia (5°N), fruit is harvested in December.

Cultivated trees were observed flowering during March-April at Maracay, Venezuela (10°N), and according to M. Figueroa (personal communication 1980) first fruits would be ready to harvest in August. Figueroa also says that they always flower and fruit at the same time each year.

In Florida at 24 to 25°N Whitman (1976) describes a January-February flowering followed by a November harvest. Flowering in the Cairns area of north Queensland is reported from July to October.

Market acceptability

Romero Castaneda (1961) reports that the fruit is well liked by inhabitants of production areas in Colombia and that Humboldt and Bonpland, after discovering it in 1801 had recommended to European governments that it be cultivated in their African and Asian colonies. However, Hodge (1960) was 'a frequent eater of these fruits during a sojourn in Colombia' and he agreed with the opinion of Popenoe (1924) that it 'cannot be considered a fruit of great value'. M. Figueroa (personal communication 1980) also does not consider it to have a lot of potential for commercial exploitation.

As mentioned previously, pulp of the fruit is fibrous but sweet and agreeable in flavour. Hodge (1960) reports that some people consider the fruit too fibrous. Perez-Arbelaez (1956) also mentions that it is excessively fibrous, but that it is quite agreeable if you have plenty of water on hand to wash it down. Its fibrous nature is also reflected in one of its common names, chupachupa, meaning chewy.

Of course market acceptability often depends on what people are accustomed to consuming and user education over time can change attitudes towards fruit acceptability.

Carle and Thompson (Cairns residents who have visited the Amazon) rate the fruit much higher in palatability than do Popenoe, Hodge and Figueroa (personal communication 1980).

Export potential

Romero Castaneda (1961) advised that as it stands up well to transport it would be worthwhile carrying out a trial to introduce it to markets outside Colombia. He suggests that Tumaco is the area of Colombia from where this could be done, because of its proximity to the sea, abundance of the tree in that area, its good growth without pest problems and its good flavour.

Whether it has in fact been introduced to markets ouside Colombia is not known.

Cultural Details

Propagation

Generally from seed. Both Perez-Arbelaez (1956) and Hodge (1960) reported that trees in culture are seedlings. However Clement (personal communication 1980) has successfully propagated selections using wedge and side-veneer grafts in Brazil.

Management

L. Muller (personal communication 1980) says they need protection from wind. No further information is available.

Irrigation

Because of its origin in lowland rainforest areas, it is presumed that irrigation would be necessary for establishment, and for mature trees in northern Australia. Amounts and frequency of application are unknown.

Fertilization

No documented information is available on fertilization. In the Northern Territory and Queensland we would recommend an N:P:K mix together with organic mulches, as all soils are deficient in these elements and organic matter.

Pests and diseases

Perez-Arbelaez (1956) reports that it is very susceptible to attacks by insect larvae with no mention of the names of insects involved or control methods. Whitman (1976) states that the Keys white fly *(Aleurodicus dispursus)* and the Cuban May beetle *(Phyllophaga bruneri)* can strip trees of foliage.

In the Cairns area the main problem is with black beetle *(Rhyparida* spp.*)* which causes severe leaf damage during the summer and autumn months. No pathogens have yet been evident in Queensland.

Harvesting methods

According to Hodge (1960) the fruit is ready for picking when a readily visible ring of lighter coloured tissue is seen where the edge of the persistent calyx meets the fruit. This is due to growth at the time of maturity which brings a bit of the fruit's surface out from under the protection of the calyx.

No further information is known.

Marketing characteristics

As mentioned above it stands up well to transport, but one characteristic detrimental to its market acceptability is its excessive fibre.

Considering its nutritional value as a market characteristic, chemical analysis of edible pulp is shown in the following table.

Analysis of 100 grams of edible pulp, *M. cordata*

Calories	49	Calcium	25 mg
Water	85.1 g	Phosphorus	32 mg
Protein	1.1 g	Iron	1.4 mg
Fat	0.1 g	Vitamin A	1000 I.U.
Carbohydrate	12.4 g	Thiamin	0.05 mg
Fibre	0.6 g	Riboflavin	0.09 mg
Ash	0.7 g	Niacin	0.4 mg
		Ascorbic Acid	20 mg

(Gongora Lopez and Young Lopez 1953)

Transport characteristics

M. cordata stands up well to transport due to its thick skin.

References

Gongora Lopez, J. and Young Lopez, N. (1953), 'Table de Composicion de Alimentos Colombianos'. Bogota. Quoted by Romero Castaneda, 1961.

Hodge, W.H. (1960), The South American Sapote. *Economic Botany* **14**, 203.

Perez-Arbelaez, E. (1956), 'Plantas Utiles de Colombia'. Libreria Colombiana Camocho Roldan Cia Ltda., Bogota, p. 720.

Popenoe, W. (1924), Economic fruit bearing plants of Ecuador. *Contributions U.S. National Herbarium* **24**, 122. Quoted by Hodge, 1960.

Romero Castaneda, R. (1961), 'Frutas Silvestres de Colombia'. Bogota, **1**, 131.

Whitman, W.F. (1976), South American sapote. *Proceedings of the Florida State Horticultural Society* **89**, 226.

Uphof, J.C.T. (1968), 'Dictionary of economic plants'. 2nd edition, Cramer, Lehre, p. 334.

Chapter 7

BURSERACEAE

PILI NUT
Canarium ovatum Engl.

by B.W. CULL

Botany

Pili nut belongs to the family Burseraceae. The genus *Canarium* is variably reported to have from 80 to 100 species of which 10 bear edible nuts.

C. ovatum is a fairly large tree growing to a height of about 20 m and a trunk diameter of 40 cm. If planted too closely they will lose their lower limbs readily and only have an upper canopy. Leaves are alternate, compound with opposite smooth leaflets, rounded at the base and pointed at the top. The leaflets are from 10 to 20 cm long.

The species is dioecious, with the flowers borne in rather large terminal clusters. The ovary contains 3 locules, each with 2 ovules but nearly always only 1 of the 6 ovules develops.

It flowers in summer (May-June in the Philippines) and matures its fruits late summer, early winter.

Fruits consist of a single, nearly cylindrical, oily, pleasantly flavoured kernel, a hard thick endocarp and a thin layer of soft pulp which has a glossy skin, dark purple when ripe. The edible kernel, with 70 to 80% oil, has a flavour resembling that of sweet almonds. The pulp of the fruit is edible but not tasty. It does contain an oil which can be extracted.

The nuts are slender, tapering to the ends and triangular in cross section. Sometimes they attain a length of 6 cm and a diameter in the middle of 2 cm. They can, however, be much blunter and shorter with the same diameter.

Armour (1966) working with a collection of 71 trees showed that fruits when mature weigh in the vicinity of 23 to 43 g. Of this 70 to 84% can be pericarp, 12 to 22% endocarp and 3 to 7% kernel. In dried nuts a figure of 25% kernel represents the peak of kernel recovery recorded.

Climatic requirements

The Philippines is the country of origin and it is recorded in S. Luzon, Polillo, Samar, Layete and Mindanao. It occurs in primary forests up to 500 m above sea level

The trees are being grown successfully in Malaysia, Hawaii and Honduras. There are no reports of plants in more subtropical to temperate areas. There is no doubt that they are very sensitive to frost.

Soil requirements

Soil requirements are not reported.

Genetic diversity

Within *C. ovatum* there appears to be certain useful diversity. Both proportional weight of kernel to total weight of fruit and also weight of dried nut vary. Selection in this respect would increase kernel yields.

Shape of nut also varies from an elongated form to a more regular shape. This variation could be of importance to cracking machinery. Thickness of shell also varies and its relationship to handling, cracking and ultimate kernel recovery could be important.

A wide range of other *Canarium* species exist. *C. luzonicum* which is often called pili nut has small nuts and has its origin in a similar location as *C. ovatum.*

C. vulgare and *C. indicum* are closely related and are called Java almonds or Kenari nut. The nuts of these once again are similar in shape to pili nuts but are smaller. *C. indicum* is found in Malaysia and New Guinea and is thought to have superior eating qualities.

Nuts of the various species *C. mehenbethene, C. salomonense, C. commune* and *C. nungi* are eaten extensively in New Guinea, the Solomons and the New Hebrides. The 'Garlip' or 'Garlap' nut is either *C. mehenbethene* or *C. indicum* or both, and is found in New Guinea and related islands. There appears to be some merit in collecting these latter species, particularly those with high kernel recovery.

C. album and *C. pimela* are Chinese in origin and in the pickled form are known as Chinese olives.

C. schweinfurthii is the main food producing species in central Africa and is known as incense tree or bush candle tree.

C. australeanum, C. baileyanum and *C. muelleri* are found in Australia, the former in Arnhem land, *C. muelleri* around the Atherton Tableland, and *C. baileyanum* extending down to Lismore through Queensland.

The actual naming and identification of species is somewhat confused with different authors quoting different species names for the same nut.

The value of this genetic diversity is unknown. Many of the species either bear poor nuts or no nuts at all. Their wide adaptability could be of importance should breeding be considered.

Production season and time of cropping

Pili nut appears to flower in early summer to midsummer maturing its crop possibly winter to early summer. This harvest season would suit Australian weather conditions.

Yield potential

Recorded average yields in Honduras for mature trees (10 years) are in the vicinity of 30 to 40 kg of nuts in shell. Quoted yields from the Philippines are greater than 100 kg from single trees.

Market acceptability

The Philippines exported 1186 tons of nuts in 1965. However, it is believed now that this trade has lapsed.

It is claimed that they have a similar or greater appeal than almonds and can be used in the same way. Roasting appears to improve the flavour and acceptability.

Cultural details

No commercial plantations appear to exist hence cultural details are not available. Of interest is the need for vegetative propagation because of the dioecious nature of the tree.

Processing and marketing characteristics

Traditionally the pulp of the fruit is removed by hand either following fermentation or soaking in water at a temperature of 40 to 50°C. Temperature higher than this can cause a premature breakdown in the nuts. Otherwise it is reported that dried nuts store well. Moist nuts go rancid within 1 to 2 days.

No mechanical cracking is reported. The kernel has a testa which can also be removed with the aid of hot water.

The flavour of the nuts can be improved by roasting. The nut has been used in confectionery with chocolate, in home cooking and as an infant food.

References

Armour, R.P. (1966), Kernel weight in relation to fruit size in the pili nut *(Canarium ovatum* Engl.*). Amer. Soc. Hort. Sci. Caribbean Reg. Proc.* 1965 (13th Meeting), **2:** 137.

Further reading

Howes, F.N. (1948), 'Nuts: their production and everyday uses'. Faber and Faber, London pp. 24, 87, Plate 3.

Leenhouts, P.W. (1959), 'A monograph of the genus *Canarium (Burseraceae)*'. Leiden, Ijdo. (repr. from *Blumea* **9,** 275).

Leenhouts, P.W. (1965), Notes on *Canarium (Burseraceae)* in the Solomon Islands. *Blumea* **8,** 163.

Yen, D.E. (1974), Arboriculture in the subsistence of Santa Cruz, Solomon Islands. *Economic Botany* **28,** 247.

Menninger, E.A. (1977), 'Edible nuts of the world'. Horticultural Books, Stuart, Florida.

Chapter 8

EBENACEAE

by A.P. GEORGE

BLACK SAPOTE
Diospyros digyna Jacq.
D. ebenaster Retz.

Botany

The Ebenaceae family comprises four genera, *Diospyros, Euclea, Royena* and *Tetraclis.* Of these genera, *Diospyros* is the largest, containing 200 species, mostly of tropical and subtropical origin (Popenoe 1920). The black sapote, *Diospyros digyna* Jacq. (synonym *D. ebenaster* Retz.) is so named because the fruit of this species, at maturity, contains a dark, chocolate brown pulp. This species appears to have originated in Mexico perhaps as early as 5000-3000 years B.C. Today, it is cultivated on a minor scale in Hawaii, Florida, California, Mexico, the West Indies and the Philippines. In Hawaii, it is referred to as the black persimmon, in Puerto Rico as guayabote, and in the Philippines as zapote negro (Popenoe 1920). The only close relative of this species, of commercial importance, is the oriental persimmon, *D. kaki.*

The most important botanical feature of this species is the olive green, oblate, ribbed or lobed fruit. The pulp which is contained within the thin skin is soft, unctuous, dark chocolate brown, and of sweet flavour but without aroma (Popenoe 1920). Berry *et al.* (1977) describe the fruit as having a sweet persimmon-like flavour and texture, with the flesh and peel as having a high tannin content. Fruit size varies from 5 to 15 cm in diameter. Trees are 6 to 10 m high.

Climatic requirements

The black sapote is adaptable to a wide range of tropical and subtropical climates. Temperature, however, is the factor limiting wider geographical distribution, with young trees being killed by temperatures below 0°C and mature trees by temperatures 2°C below zero (Fisch 1975; Campbell, Knight Jr. and Zareski 1977).

Soil requirements

The species appears capable of growing in a wide range of soil types from sandy loams to light clays.

Genetic diversity

Very few improved selections of black sapote have been made and the range of genetic diversity from which further selections can be made is now known. Suitable rootstock species or varieties have not been evaluated.

Production season and time of cropping

In Florida, this species fruits between January and May with the heaviest production occurring between January and February (Green 1980). In Queensland, the fruiting period would most likely coincide with our autumn and early winter period.

Yield potential

In a comparison of tree productivity of several tropical and subtropical tree crop species, Green (1980) rates the black sapote as a high producer. However, exact yield data are not presented.

Export potential

The development of export markets for the fresh fruit product is feasible because of the long cool storage life which can be achieved. The fruit may also be processed into a wide range of products which could also be easily exported.

Cultural details

Propagation

The black sapote can be propagated by budding or grafting. Wester (1924) describes an early method for shield budding. A more recent propagation technique is described by Clift (1980). This technique involves grafting of leafy scions of tender tips under mist.

Management

Little is known about cultural methods suitable for the black sapote. Young trees are planted out when 30 to 60 cm high with tree spacing varying between 8 to 14 m depending on soil type and environment. Some frost protection is needed with young trees up to 3 years old. Little or no information is available on training, pruning or fertilizing.

Irrigation

Irrigation is regarded as essential for maximum production (Fisch 1975); however, rates and timing have not been defined.

Harvesting methods

The fruit should be picked when mature and allowed to ripen in a dry, cool place. The fruit can be hard one day and ready to use 24 hours later. The ripe fruit is dull, olive green and is as soft as a marshmallow (Knight Jr 1976). Fruit flavour is similar to persimmon

Marketing characteristics

The fruit is picked when firm and mature and can be sent to distant markets. However, once the fruit begins to soften, usually within 3 to 6 days after picking, handling becomes difficult The use of cool storage extends the marketing life considerably.

Storage characteristics

The black sapote can be cool stored for considerable periods of time. Studies by Berry *et al.* (1977) have indicated that green, unripe fruit can be held at 10°C for several months and will soften and ripen within 48 hours after warming to 29.4°C. These researchers also found that a close correlation existed between the decrease in fruit circumference with ripening and the degrees Brix, and that this relationship could be used as an index of ripeness.

Processing characteristics

The fruit may be eaten fresh or used in ices. The taste is usually improved by blending it with orange, lemon or lime juice. It may be cooked with lemon or made into preserves. In some areas, the fruit is mixed with wine, sugar and cinnamon to make a delicious dessert. In Mexico, the ripe fruit is made into preserves which are reported to be of excellent quality. Brandy is also said to have been made from the pulp (Fisch 1975).

References

Berry, R.E., Wagner Jr, C.J., Shaw P.E. and Knight Jr, R.J. (1977), Promising products from tropical fruits. *Food Product Development* **11**, 109.

Campbell, C.W., Knight Jr, R.J. and Zareski, N.L. (1977), Freeze damage to tropical fruits in southern Florida in 1977. *Proceedings of the Florida State Horticultural Society* **90**, 254.

Clift, C. (1980), The 'Bernecker' black sapote, *in* Newsletter (April), The Rare Fruits Council International Inc., Miami, Florida.

Fisch, B.E. (1975), The black sapote. *California Rare Fruit Growers Yearbook* **7**, 91.

Green, J. (1980), Fruiting calendar for south Florida, *in* Yearbook, The Rare Fruits Council International Inc., Miami, Florida, p. 86.

Knight, Jr, R.J. (1976), The black sapote *in* M.W. Richardson (ed.), 'Tropical fruit recipes'. The Rare Fruits Council International Inc., Miami, Florida.

Popenoe, W. (1920), The black sapote *in* 'Manual of tropical and sub-tropical fruits'. Hafner Press, New York.

Wester, P.J. (1924), Food plants of the Philippines. Bulletin 39, Philippine Department of Agriculture and Natural Resources, Bureau of Agriculture, Manila.

Chapter 9

EUPHORBIACEAE

by P.B. SCHOLEFIELD

The Euphorbiaceae family includes few species that could be described as having commercial horticultural potential.

Botany

BIGNAY, BUNI, WUNI
Antidemsa bunius (L) K. Spreng.
An evergreen tree 6 to 10 m high with dark green, glossy leaves (7 to 15 cm x 4 to 6 cm). Dioecious with many small flowers borne in racemes 5 to 15 cm long. The small, red fruits are packed in tight bunches and become dark red to black when ripe.

HERBERT RIVER CHERRY, WOOLMI
Antidesma dallachyanum Baill.
A small dioecious, evergreen tree 6 to 8 m high found along the fringes of rainforest in coastal north Queensland. Deep green leaves (5 to 10 cm long) are thick and leathery. Small greenish flowers are borne in racemes giving rise to clusters of fruits (12 to 25 mm in diameter) with colours varying from cream with a red blush to dark purple.

Baccaurea spp.
Burkill (1966) lists 19 species of the genus *Baccaurea*, but few have useful fruits (*B. dulcis* Muell. Arg. RAMBAI, TJUPA; *B. motleyana* Muell. Arg. MENTENG NEGERI, RAMBAI; *B. racemosa* Muell. Arg. KEPUNDUNG; *B. sapida* Muell. Arg.). Fruits are 20 to 35 mm in diameter, green to pink-red and with acid flesh.

OTAHEITE GOOSEBERRY, STAR GOOSEBERRY, CHERMAI
Phyllanthus acidus (L) Skeels.
A tree 6 to 8 m high with opposite, pale green leaves arranged along thin shoots giving them the appearance of a compound leaf. The tree is evergreen but may appear deciduous under water stress. Small pinkish flowers are borne in clusters originating from woody stems of the tree. The fruit is roundish, slightly ribbed and 10 to 30 mm in diameter with a light yellow colour and acid taste.

EMBLIC, MELAKA, MYROBALAN
Phyllanthus emblica (L)
Large deciduous tree (15 m high) with distinctive thick, grey bark which flakes off in patches exposing white bark underneath. Leaves are leathery and fern-like. Small flowers are borne in dense clusters under the terminal leaves of shoots and give rise to rounded fruits up to 30 mm in diameter.

Climatic requirements

Tropical to warm subtropical climate is preferred.

Soil requirements

The *Phyllanthus* species are known to prefer well-drained soils but little information is available on the other species.

Genetic diversity

Very little serious selection has been done on any of these species and great genetic diversity is exhibited.

Production season and time of cropping

A. dallachyanum flowers in the wet season and the fruit mature from April to July. Cropping time of other species is not well defined.

Yield potential

A. dallachyanum and *P. acidus* are heavy bearers.

Market acceptability

Most of the fruits in this family are small, and in some cases, very acid with thin flesh. The common uses reported are jams, jellies, pickles and chutneys. The species are therefore likely to be more suitable as 'backyard' trees.

Export potential

The export potential of the fruit is low.

Cultural details

Propagation is by seed or cuttings.

Harvesting methods

Because most of the fruits are small, hand harvesting would be expensive. Some form of mechanical tree or limb shaker may prove suitable, particularly as the fruits need to be processed.

References

Burkill, I.H. (1966). Dictionary of the economic products of the Malay Peninsula, volume 1, *Antidesma*, p. 185; *Baccaurea*, p. 279. Ministry of Agriculture and Cooperatives, Kuala Lumpur.

Further reading

Bentham, G. (1973). 'Flora australiensis'. L. Reeve and Co., Covent Gardens, volume VI, 84.

Chapter 10

FLACOURTIACEAE

by M. SEDGLEY

GOVERNOR'S PLUM
Flacourtia indica (Burm. f.) Merrill
Flacourtia ramontchi L'Her.; *F. sepiaria* Roxb.

RUKAM
Flacourtia rukam Zoll. & Mor.

LOVI-LOVI
Flacourtia inermis Roxb.

PANIALA
Flacourtia jangomas (Lour.) Rauschel

CEYLON GOOSEBERRY, KITEMBILLA
Dovyalis hebecarpa (G. Gardn.) Warb.
Aberia gardneri Clos.

AFRICAN GOOSEBERRY, KOSHUM
Dovyalis abyssinica (A. Rich.) Warb.

FLORIDA GOOSEBERRY
Dovyalis abyssinica x **hebecarpa**

KEI APPLE
Dovyalis caffra (Hook. f. & Harv.) Warb.
Aberia caffra Hook, f. & Harv.

The species, native to Asia and Africa, will be treated as a group as they appear to be very similar in most aspects. Also there is sexual compatability between some *Dovyalis* species and graft compatibility amongst the *Flacourtia* species. There are no established superior cultivars as yet and these will probably be developed by interspecific crossing and grafting.

Botany

All are shrubs or small trees up to 15 m high. *Dovyalis* is an evergreen genus whereas *Flacourtia* is deciduous. All are spiny plants with entire ovate green leaves up to 9 cm long with crenated or slightly serrated edges. All species are dioecious although occasional perfect flowered specimens exist. The flowers, produced in axillary racemes, are unisexual, pale yellow, small, and the female flowers produce an edible berry. The subglobose fruit has numerous seeds and grows up to 5 cm in diameter. The fruit of *Flacourtia* species is dark red whereas the fruit of *Dovyalis* species is yellow with the exception of *D. hebecarpa* which has a purple fruit. The fruit may be hairy or smooth depending on the species.

Climatic requirements

All species require warm subtropical conditions. They are all damaged by frost although *D. caffra, D. abyssinica* x *hebecarpa* and *F. indica* are moderately hardy to –3.5°C. They will withstand dry conditions.

Soil requirements

All species will tolerate a wide range of soil types, preferring a pH range of 5.5 to 8.0. However, *D. abyssinica* x *hebecarpa*, the species showing the most promise as a crop plant, thrives best in Florida on a rich soil of pH 5.5 to 6.5

Genetic diversity

Variation in the fruit, from sweet to astringent, has been reported in *F. indica*, *F. inermis* and *D. abyssinica* x *hebecarpa*, and there is variation in spininess in *F. indica*. Although all species are dioecious, occasional monoecious plants or plants with some perfect flowers occur in *F. indica*, *D. hebecarpa* and *D. abyssinica* x *hebecarpa*. *D. caffra* bears after 4 to 5 years so breeding and selection could probably proceed quite rapidly. There are no established cultivars of any species.

Production season and time of cropping

Enormous variation is reported in the time of flowering and time of cropping, within species, between species and from location to location. The reports suggest that the flowering and fruiting seasons are protracted and dependent largely upon local conditions. Under Florida conditions, *D. abyssinica* x *hebecarpa* flowers from September to January and fruits from December to April.

Yield potential

All species are reported to bear heavily.

Market acceptability

Existing selections have acceptable fruit colour, and if cultivars with large, sweet fruit could be developed then the fruit would probably be acceptable for the fresh market. Existing overseas selections are mainly suited for presevation (Knight 1974). *D. hebecarpa* is reported to be high in vitamin C.

Export potential

There is no overseas information on export potential, although export of the processed product may be feasible.

Cultural details

Propagation

All species can be propagated by the budding of seedlings. Air layering, cuttings, and suckers are also successful in most cases tested. There is no graft incompatability between *F. rukam*, *F. inermis* and *F. indica*, and *D. abyssinica* x *hebecarpa* can be grafted onto *D. hebecarpa*.

Management

Great variations in cultural requirements are reported for the different species (Maxwell 1967). As there are no established cultivars these reports probably largely reflect plant to plant variation. Approximately 8 to 12 m spacings with some pruning of older trees is suggested for *D. abyssinica* x *hebecarpa*. As the species are dioecious, the interplanting of 1 male plant to 15 females has been quoted. However, the selection of bisexual cultivars should overcome this requirement.

Irrigation
For maximum production irrigation is needed particularly during fruiting. *D. caffra* is reported to be the most drought tolerant of the group.

Pests and diseases
Mediterranean fruit fly *(Ceratitis capitata)* attacks *D. caffra* and *F. indica* fruit, and the lepidopteran *Icthyura restitura* defoliates *F. inermis* in India (Abraham and Remamony 1978). The West Indian red scale *(Selenaspidus articulatus)* has been reported on *D. caffra* in the Nairobi area, but can be controlled biologically by parasitism with *Aphytis roseni* and *Habrolepis rouxi* (Greathead 1976).

Storage and processing characteristics

There is no information available but the fruit of all species can be preserved as jams, jellies and sauces.

References
Abraham, G.C. and Remamony, K.S. (1978), Occurrence of *Ichthyura restitura* Walker (Notodontidae: Lepidoptera) as a serious pest of Lovi-Lovi, *Flacourtia inermis* Roxb. *Agr. Res. J. Kerala* **16,** 79.
Greathead, D.J. (1976), An east African parasite *(Aphytis roseni* Debach and Gordh, Aphelinidae) for the control of *Selenaspidus articulatus* Mogen (Diaspididae) in Peru. *East African Agricultural and Forestry Journal* **41,** 349.
Knight, R.J. Jr (1974), Special problems in tropical fruit production. XIXth International Horticultural Congress, Warsaw.
Maxwell, L.S. (1967), 'Florida fruit'. Maxwell, Florida.
Further reading
Kennard, W.C. and Winters, H.F. (1960), Some fruits and nuts for the tropics. USDA Miscellaneous Publication 801.

Chapter 11

GUTTIFERAE

MANGOSTEEN by D. McE. ALEXANDER
Garcinia mangostana L.

MAMMEY
Mammea americana L.

Of the 35 genera in the Guttiferae family, 8 of them namely, *Allanblackia, Garcinia, Haronga, Mammea, Pentadesma, Platonia, Rheedia,* and *Symphonia* produce edible food.

Oil from seeds of *Allanblackia oleifera* Oliv. and *A. stuhlmanii* Engl. is used locally in central and east Africa as butter and for cooking oil. Uphof (1959) lists 38 edible species of *Garcinia* which are native to an area extending from the Philippines to Queensland and westward through Indo-China to Bengal and Sumatra. Fruits and seeds of *Haronga madagascariensis* Chois. are used in east Africa and Madagascar. *Pentadesma butyracea* Sabine produces seeds which are a source of edible fat. From the Americas *Mammea americana, Platonia insignis* Mart, and the *Rheedia* species, *brasiliensis* Planch and Triana, *edulis* Planch and Triana, *macrophylla* Planch and Triana, and *madruno* H.B.K. all produce edible fruits. *Symphonia macrophylla* Baker from Madagascar also produces edible fruits.

The outstanding fruit of this family is produced by *Garcinia mangostana* but several others including *Mammea americana, Garcinia tinctoria* (D.C.) W.R. Wright, *G. penduncula*ta Roxb, *G. multiflora* Champ., *P. insignis* Mart, *R. brasiliensis* Planch and Triana, *R. macrophylla* Planch and Triana, and *R. madruno* H.B.K. are worthy of cultivation.

Garcinia is a large genus of tropical evergreen trees or shrubs from tropical Asia. Many bear edible fruits ranging in size from 1 to 2 cm to 10 to 12 cm in diameter with thin to thick woody skins. The 50 species of *Mammea* are represented in the American tropics, India, Malaysia, Madagascar and the Pacific Islands. The American species, *Mammea americana,* probably has the best fruit. *Rheedia* species are native to central and South America and the West Indies. *R. madruno* of Ecuador has the largest fruit (6 to 7.5 cm) and a flavour resembling mangosteen. Fruits of *R. brasiliensis,* the bakupari, which is indigenous to the state of Rio de Janeiro in south-western Brazil resemble those of *Platonia insignis,* the bakuri from northern Brazil, but are smaller (4 cm long) and not considered so delicious.

MANGOSTEEN
Garcinia mangostana L.

Botany

G. mangostana, the mangosteen (manggis, Malay), is a small, slow growing, pyrimidal shaped evergreen tree, up to 10 m high which has opposite, unifoliate, thick leather leaves (15 to 25 cm x 7 to 13 cm).

It is dioecious but female trees produce fruit with apomictic seeds when grown in isolation. Male trees are rare to non-existent. When damaged all plant parts excrete a yellow latex. Flowers are white to cream-coloured and terminal with 4 sepals, 4 petals and a globose ovary with a thick stigma divided into 4 to 8 lobes. Fruits have a thick reddish purple shell and are segmented with 4 to 8 segments of which 1 or 2 containing seeds are larger than the rest. The seeds are viable but are short lived. Seedlings commence fruiting when trees are 6 to 15 years old.

Climatic requirements

An equatorial climate, a high even temperature with high humidity, is required. Trees are killed by temperature down to 5°C and growth is slow below 20°C. Leaves are scorched at temperatures above about 35°C. Trees can survive a dry period provided they are kept well irrigated.

Soil requirements

The mangosteen will grow on a wide range of soils provided the water regime is satisfactory. A slightly acidic, clayey soil high in organic matter with a moving water-table at about 2 m is preferred.

Genetic diversity

Because seeds are nucellar there is little diversity but Burkill (1966) has reported a race in the Sulu Islands with thicker rind and more acid flesh while Almeyda and Martin (1976) have reported a variety in Borneo with 4 segments and 4 seeds. In orchards in Panama and Honduras large differences in yields from tree to tree have been observed.

Production season

In many regions two crops are produced each year depending on growing conditions and number of growth flushes. In Sri Lanka the main season for low areas is May and July and at higher elevations September to October. In Puerto Rico unshaded trees produce fruit in July and August while trees grown in shade fruit in November and December.

Yield potential

Mature trees yield 500 to 1500 fruits.

Market acceptability

Mangosteen are excellent as fresh fruit. Fruits can be preserved when immature or dried when mature-ripe. For market, fruits must be picked at early maturity but if harvested too soon fruits do not develop full flavour. Fruit harvested at full maturity readily spoil.

Cultural requirements

Propagation

Propagation is by seed or grafting onto seedlings of *G. mangostana, G. tinctoria, G. cochinchinsis* or some other species or relative. Node to node wedge grafting has proved most successful but budding is sometimes successful.

Management

Young trees require shading. All trees require a constant supply of water and will withstand some waterlogging.

Irrigation
Irrigation is needed every 2 weeks during the dry season. Soil must be kept from drying out particularly in the early years.

Fertilization
Mature trees require about 7 kg of complete (10:10:9) fertilizer per year.

Pests and diseases.
Gamboge canker affects fruit. Yellow latex from damaged fruits may seep into flesh giving a bitter taste.

Harvesting methods

Mature fruits drop. For marketing it is best to hand harvest before fruits are fully ripe.

Marketing characteristics

Fresh fruit can be marketed up to 21 days after harvest.

Transport characteristics

Because of its thick shell fruits transport readily.

Processing characteristics

Immature fruit can be canned. Mature fruit tends to lose flavour on canning but can be dried successfully.

MAMMEY
Mammea americana L.

Botany

Mammey is a large evergreen tree up to 25 cm with glossy, tough, leathery leaves (10 to 20 cm x 5 to 10 cm). Flowers are solitary or clustered in the axils of young shoots, 2 sepals, 4 to 6 white, fragrant petals, stamens numerous, ovary 2 to 4 celled with 1 or 2 ovules in each cell. Fruit is globose or oblate 7.5 to 15 cm in diameter with a rough, brown, thick, pliable skin and firm orange or yellow flesh. Each fruit has 1 to 4 large seeds.

Climatic requirements

Mammey favours hot humid regions below 1000 m. Temperatures below 5°C are often fatal. Preferred temperature range is 25 to 35°C.

Soil requirements

A deep, well-drained sandy loam rich in organic matter is preferred.

Genetic diversity

Variations have been observed in fruit size, quality and bearing capacity of seedling trees.

Production season and time of cropping

Cropping is during the summer months in the West Indies. Trees are slow to come into cropping.

Market acceptability

Fruit has a pleasant flavour, resembling apricot and is excellent when stewed or made into jams or jellies. Immature fruits have a high pectin content and can be used to gel other fruits.

Cultural requirements

Propagation

Propagation is by seed or it can be grafted or budded onto seedlings.

Management

Young trees require shade. Spacing about 18 m.

Irrigation

Soil needs to be kept moist.

References

Almeyda, N. and Martin, F.W. (1976), Cultivation of neglected tropical fruits with promise. Part 1. The mangosteen. USDA, ARS-S-155.

Burkill, I.H. (1966), 'A dictionary of the economic products of the Malay Peninsula'. Min. of Agric. and Cooperatives, Kuala Lumpur.

Uphof, J.C. (1959), 'Dictionary of economic plants'. Weinheim, Germany.

Further reading

Watson, B.J. (1978), An appreciation of five tropical fruit tree species and their potential in Queensland. Queensland Department of Primary Industries Thesis.

Chapter 12

LAURACEAE

by A.W. WHILEY

AVOCADO
Persea americana Mill.

The avocado is indigenous to the mainland of tropical America and has had a long history of cultivation by the native people of these countries. Its value as a food source is recorded in the Aztec and Inca civilizations and today it still ranks high in the diet of the indigenous population.

The English word avocado is a corruption of the Spanish 'ahuacate' derived from the Aztec word 'ahuacatl'. There are several other names the fruit goes under in the countries of origin but it is now universally known as the avocado throughout its adopted production centres.

The avocado was introduced in Australia by the Kamerunga State Nursery in Queensland late in the nineteenth century but it was not until the 1930s that named varieties were imported and grown. Traditionally, south-east Queensland and northern New South Wales have been the major production centres but in more recent times the area has expanded to the Atherton Tablelands in north Queensland and the irrigated settlements along the Murray River in New South Wales, Victoria and South Australia.

Botany

The avocado belongs to the genus *Persea* which is in the predominantly tropical laurel family (Lauraceae) with the trees that produce camphor and cinnamon. There are some 50 species of *Persea* (Chandler 1958) and nearly all are of American origin.

Botanical classification of the avocado has had a diverse history with two species being initially recognized. However, recent authorities now recognize only one species, *Persea americana*. This species can be usefully subdivided into three horticultural races: Mexican, Guatemalan and West Indian, named after their presumed centres of origin.

The avocado grows into a large, dome-shaped, evergreen tree with some specimens attaining a height of 18 metres. The leaves are simple and entire and multiform in shape ranging from lanceolate to ovate. The length ranges between 100 mm to 400 mm.

The small, pale green, perfect flowers are grouped in a compound panicle of racemes. Most flowering shoots are terminal to the last vegetative flush. A tree may produce a million or so flowers during one bloom period but carry only 0.1% through to mature fruit.

The avocado has a unique flower behaviour which can be termed 'protogynous, diurnally synchronous dichogamy'. The dichogamy is protogynous as in each flower the pistil matures before the stamens. The flowering behaviour is synchronous in that all open flowers on a tree of any given variety in a similar environment are female at one time and male at a different time. The synchronization is diurnal, for each tree is functionally female one part of the day and functionally male during another part of the same day.

Finally, the flowering patterns of trees can be classified as 'A' or 'B' types. All trees of the same variety react identically. A variety is classified as 'A' type if each flower is functionally female in the morning and functionally male the following afternoon; or 'B' type if each flower is female in the afternoon and then male the following morning (see table 1).

Table 1. A schematic presentation of 'A' and 'B' type flowering patterns in the avocado

	Day one		Day two	
Type	a.m.	p.m.	a.m.	p.m.
'A'	Female	Closed	Closed	Male
'B'	Closed	Female	Male	Closed

The flowering behaviour outlined above is regular only under conditions of warm weather. Bergh (1969) reports day and night minimums as 25°C and 15°C respectively while Sedgley (1977) largely supports this with results from controlled studies, also indicating the most suitable temperature regime for floral behaviour at 25°C and 15°C day and night respectively.

Fruits of the avocado have a single large seed, made up chiefly of two cotyledons, surrounded by a thick fleshy mesocarp which is covered by a skin of varying thickness.

The fruits of different varieties vary greatly in size (from less than 1 kg to over 3 kg), while the form is commonly pear-shaped or oval but ranges from round to oblate. The colour varies from yellow-green to almost yellow through many shades of green to crimson, maroon, brown, purple and almost black. The pulp oil content varies from about 3% to over 30%.

Climatic requirements

While basically a subtropical plant, the avocado shows adaptability to a wide range of environments as can be seen by scanning world production centres. To some extent this range can be attributed to the genetic diversity of the three horticltural races; at one end of the scale there is the West Indian race which prefers the warmer conditions of the tropics while at the other end is the Mexican race which can withstand –5°C for short periods. The Guatemalan race falls between these two as far as temperature tolerance is concerned.

In general the optimum growing temperatures are around 25 to 30°C day and 15 to 20°C night while a cooler rest period during the winter (approximately 10°C lower) will stimulate flower production. Established trees will tolerate temperatures to 40°C but prolonged exposure above this temperature with low relative humidity results in stress and loss of productivity.

The avocado has moderate requirements for moisture whether it be from natural rainfall or irrigation. In practically all commercial production centres irrigation supplements natural rainfall which tends to be irregular. Critical periods of moisture demand are during flowering and fruit set, 3 months later when crop adjustment occurs, and during the following period of fruit filling and maturation. Trees are very susceptible to waterlogging. Even a short period with anaerobic conditions around the roots can lead to loss of roots and may even kill the tree. Such conditions are also favourable to the root rotting fungus *Phytophthora cinnamomi.*

The tree is subject to wind damage and prefers a sheltered position for optimum growth.

Soil requirements

Contrary to common belief, the avocado will grow successfully on a wide range of soil types. However, it is the drainage feature of the soil in relation to natural precipitaion which becomes the dominant criterion as to the success of that soil type.

Some avocado cultivars

Variety	*Race	Flower type	Tree growth	Fruit shape	Skin colour	Skin thickness	Maturity time	Relative seed size	Eating quality	Comments
Hass	G	A	medium spreading	ovate	black	medium thick	late	small to medium	excellent	Sensitive to frost and heat. Good regular bearing. Fruit holds well.
Fuerte	M x G	B	large spreading	pear	green	medium	early	medium	excellent	Tree has moderate cold tolerance but sensitive to temperature for fruit set. Prone to anthracnose.
Sharwil	G x M	B	large spreading	pear	green	medium thick	mid	small	excellent	Sensitive to frost. Performs best in subtropics.
Rincon	M x G	A	compact weeping	pear	green	medium thin	early-mid	large	good	Sensitive to frost. Produces heavy crops but bears biennially with age.
Wurtz	G	A	compact weeping	pear	green	medium thick	late	large	fair	Sensitive to cold. Produces heavy crops but bears biennially with age.
Edranol	G	B	medium upright	pear	green	thick	mid	medium	excellent	Prefers warm conditions. Slow to produce but bears heavy crops. Biennial tendency.
Hazzard	G	A	medium upright	pear	green	thick	mid-late	medium	good	Sensitive to frost. Slow to produce. Fruit has short tree life and is prone to splitting.
Zutano	M x G	B	tall upright	pear	green	thin	very early	medium large	poor	Cold tolerant precocious variety. Susceptible to pest and disease.
Bacon	M x G	B	tall upright	ovate	green	thin	very early	medium	poor	Cold tolerant. Susceptible to pest and disease.
Reed	G	A	tall upright	ovate to round	green	thick	very late	medium	good	Sensitive to cold. Precocious variety bearing consistently heavy crops.
**Pinkerton	G x M	A	medium spreading	pear	green	medium	early-mid	small	good	Cold tolerant, precocious variety bearing consistently heavy crops. Short tree life for fruit.
**Jim	M x G	—	medium upright	pear	green	medium thin	early	medium	good	Cold tolerant precocious variety.
**Santana	M x G	—	tall upright	pear	green	medium thin	very early	medium	good	Cold tolerant. Low yields in California.

* M is Mexican race; G is Guatemalan race.

** These varieties are not yet proven in Australasia.

In Australia avocado production is located largely on the northern New South Wales and southern Queensland coasts. Both areas are subject to periods of heavy rainfall due to summer cyclonic influences and soil types selected must have drainage features sufficient to prevent waterlogging for all but very short periods. Generally it is considered that 1 to 2 m of free draining soil overlying a porous subsurface is necessary if prolonged tree life is to be expected. In areas of lower rainfall such as the Sunraysia district (near Mildura in north-west Victoria), it may be possible to grow trees successfully on poorer structured soils.

In Florida, California and Israel trees are grown on calcareous soils having naturally high pH values. These soil types generally induce trace element deficiencies, particularly zinc and iron.

Genetic diversity

There is immense diversity within the *P. americana* species which can be categorized by the recognition of the three horticultural races: Mexican, Guatemalan and West Indian. Within these three races there is a range of cold tolerance. The Mexican race is tolerant to the cooler range of a subtropical climate and at the other end the West Indian race prefers a tropical climate. The West Indian race shows greater tolerance to salinity while the Mexican race is least tolerant. The West Indian fruit has a low oil content while the Mexican lines have a high oil content. The Guatemalan race is between the other two in its ranges. The Mexican race has thin skinned fruit with a dark green to purple colour while the Guatemalan race has a thick skin of pale to dark green. The West Indian race has a thin leathery skin with light green through to reddish tonings.

There are no known sterility barriers among the three races hence hybridization occurs readily. All major varieties grown in Australia are hybrids, predominantly between the Mexican and Guatemalan races.

Production season and time of cropping

Seedling avocados vary substantially in the time they take to start bearing fruit. In general the Mexican race is the most precocious but even seedling trees of this race may take 6 to 8 years to produce fruit. Some seedling trees have been known to remain barren for their lifetime.

Commercial production is based on grafted trees which guarantee fruitfulness as well as show greater precosity than seedlings. The different varieties vary in the time of fruiting but the earliest start at 2 years with most bearing by the fifth year.

Flower initiation usually occurs in late autumn on the hardened summer flushes with flowering beginning in late winter through to the spring depending on the variety. Fruit maturation occurs from 6 to 14 months after flowering depending on the variety and temperature.

In California avocados are harvested 12 months of the year (Anon. 1976; Gustafson 1976). This is achieved by use of a range of varieties grown in centres with different climatic environments. The environmental range of Florida is much narrower and the production season only lasts 6 months (Popenoe 1920).

In Australia the harvest season comes close to 12 months of the year though supply is light during January, February and March which are potentially the periods of peak demand.

At present the earliest maturing fruit comes from the Atherton Tablelands in north Queensland which is approximately 2 weeks earlier than the south-east Queensland coastal districts (Whiley and Brown 1980). By using a range of varieties the season in any one district may be extended to 7 to 8 months. However, the geographical distribution of producing areas within Australia ensures a much longer season; for example Atherton Tableland fruit of the same variety matures 4 to 5 months earlier than Sunraysia fruit.

Yield potential

The cropping characteristics of avocados can be erratic in some areas (Clarke and Clarke 1926; Bergh 1961; Gustafson 1976). Peterson (1956) correlated low temperatures with poor fruit set and further work has demonstrated the sensitivity of the flowering mechanism to temperature (Sedgley 1977).

Biennial bearing occurs to varying degrees in most areas with some varieties more prone than others (Hodgson 1947; Zerner and Gottreich 1975).

The climate in Queensland and northern New South Wales appears more favourable for fruit set than that of many other producing centres. By world standards crops tend to be more regular and higher yielding. Cropping in the Sunraysia district tends to be erratic and often upset by high temperatures or hot winds during flowering and early fruit development.

Five-year-old orchard trees average 80 kg per tree in coastal south-east Queensland while 15-year-old trees average 250 kg per tree. Individual trees have produced 500 kg of fruit in 1 year.

Market acceptability

The avocado has a unique flavour amongst fruits, consisting neither of sweetness nor acid but rather a mild 'nutty' sensation from a pulp of rich buttery consistency. Most people need to acquire a taste for the fruit over a period of time but once known the fruit is keenly sought after. It has a wide range of uses from savoury to dessert.

Markets need to be supplied with mature fruit of known quality rather than seedling types or immature fruit which quickly dampen demand.

Export potential

Currently Australia has one of the highest priced internal markets for avocados and there is little incentive to develop export outlets.

At present the only significant non-producing market is Europe which is supplied 12 months of the year from Israel and South Africa. Australian production would coincide with South African and it is unlikely that we could be competitive, as South Africa has the advantage of being closer to Europe and has lower production costs.

The Asian market has not yet been developed.

Cultural details

Propagation

Avocado trees are usually grafted to maintain varietal lines. Seedling rootstocks are commonly used though there is a trend in California to vegetatively propagate rootstocks to maintain disease resistance (Overby 1978).

In Australia seed is sown in winter and seedlings are ready to graft during late spring and summer. Cleft or whip grafting are the most common techniques with an experienced propagator achieving 80 to 95% success. Propagation is usually carried out under 50% shade as young plants sunburn very easily during the summer months.

Trees are usually ready for sale in autumn, approximately 9 months after planting the seed.

Management

Newly planted trees respond to shelter during their first year when individual tree guards are built. In some locations it may be necessary to completely cover the tree during its first winter to protect against frost damage.

Trees are usually grown with a grass sward and heavy mulching around the root zone. However, Pegg (1976) recommends an intensive cover cropping system between trees during the early years of the plantation to assist in *Phytophthora* root rot control. It is recommended that orchards be cropped with the legume *Lablab purpureus* and a forage sorghum during summer and New Zealand blue lupin during winter with crop residues slashed and lightly disced into the soil.

Tree spacing varies, depending on variety and local growth rates. In California, Platt *et al.* (1970) recommend initial plant spacing at 4.8 m x 6 m with progressive thinning as the orchard grows until a spacing of 9.6 m x 12 m is reached. Under Queensland conditions this spacing has been found to be too close as trees are crowding by the time they commence bearing. A grid of 6 m x 7.5 m is satisfactory for taller growing varieties such as Hass with eventual thinning to 12 m x 15 m but 9 m x 12 m is more suitable for spreading trees such as Fuerte which can be thinned to 12 m x 18 m.

Irrigation

Production centres with low rainfall require irrigation for successful production of avocados. In Israel Lahav and Kalmar (1977) determined that trees should be irrigated every 21 days from fruit set, and through the dry summer months. California has similar frequency requirements. In northern New South Wales and Queensland irrigation is necessary during spring and early summer if excessive fruit drop is to be avoided. Generally summer and autumn rainfall are adequate without additional irrigation. The Sunraysia district in southern Australia must irrigate regularly as rainfall is low. Salinity is a problem in this area.

Fertilization

Fertilizing practices vary with locality as soil type and rainfall greatly influence the requirements of the tree. Goodall *et al.* (1965) have established critical leaf nutrient levels for the avocado which can be used to regulate nutrient applications (see table 3).

Table 3. Critical leaf nutrient levels for the avocado — two months prior to flowering.

Element	Ranges		
	Deficient	Adequate	Excess
Nitrogen	1.6%	1.6–2.0%	2.0%
Phosphorus	0.05%	0.08–0.25%	0.3%
Potassium	0.35%	0.75–2.0%	3.0%
Calcium	0.50%	1.0–3.0%	4.0%
Magnesium	0.15%	0.25–0.80%	1.0%
Sulphur	0.05%	0.20–0.60%	1.0%
Boron	10–20 ppm	50–100ppm	100–250 ppm
Iron	20–40 ppm	50–200 ppm	—
Manganese	10–15 ppm	30–500 ppm	1000 ppm
Zinc	10–20 ppm	30–150 ppm	300 ppm
Copper	2–3 ppm	5–15 ppm	25 ppm

Nitrogen is the most important nutrient in determining yield in the avocado. Excessively high leaf levels produce vegetative growth rather than flowering. It takes careful management to adjust the nitrogen leaf level into the correct range.

The other nutrients are relatively easily managed. However, there can be problems with boron, iron and zinc deficiencies on many soil types.

Pests and diseases

The major disease of avocados is root rot caused by *Phytophthora cinnamomi.* This disease is present in most countries causing severe problems in avocado orchards. Other diseases particularly relevant to Australia are anthracnose *(Glomerella cingulata* var. *minor)* which causes fruit rots, verticillium wilt *(Verticillium dahliae)* which attacks root systems but is of minor importance, and sun-blotch caused by a viroid which distorts fruit and stunts tree growth.

Insect pests attacking avocado fruit are fruit fly, spotting bug and scale. All these may be controlled by spray programmes.

Harvesting methods

Fruit is gathered by hand from the trees once it has reached maturity. Several harvests may be made removing larger fruit first thus allowing smaller fruit to size. As trees grow larger picking aids are employed including ladders, hand operated extension cutters and hydraulic cherry pickers.

Avocados can be handled in bulk bins and fruit must be covered from sunlight once picked as heated fruit will ripen unevenly.

Marketing characteristics

The total dry matter in the edible portion of the avocado, at 30%, and the protein content at around 2%, is greater than in any other fresh fruit.

The composition of a Fuerte fruit is as follows (Popenoe 1920): water 60%; protein 2%; fat 30%; carbohydrates 7%; ash 1%.

The flesh recovery from the fruit is variable beween varieties but ranges between 70 and 80%.

Storage characteristics

On-tree storage is possible with the avocado as it will not ripen while on the tree but remains in a mature, hard condition. Blumenfeld and Gazit (1974) reported that Fuerte fruits in Israel, sometimes with germinating seeds, have stayed firm and continued to grow and accumulate oil for 10 months after attaining horticultural maturity. This phenomenon is attributed to inhibiting substances transmitted from the tree.

Post-harvest temperature is a significant factor in the time taken for avocados to ripen. Storage at reduced temperatures, for example 10 to 18°C, may extend shelf life by up to 3 weeks but fruit will ultimately soften (Chaplin 1977). Safe storage temperatures vary with variety and stage of maturity (oil content). In general later maturing varieties can be safely stored at lower temperatures than early maturing varieties in the range 2 to 8°C.

The post-harvest storage life of avocados may also be increased by sealing fruit inside a polythene bag together with a suitable ethylene absorbent. Chaplin (1977) showed that Hass fruits with ethylene absorbent in bags were still hard green after 30 days at 20°C.

Kosiyachinda and Young (1976) found that edible ripe fruit can have its life extended by up to 40 days by storage at 2°C. Fruit is only sensitive to chilling before softening occurs.

Transport characteristics

The feature of mature-hard fruit makes the avocado suitable for transporting over long distances. However, care must be taken to produce fruit free of disease as latent infections of anthracnose can cause significant losses.

Fruit are generally packed in fibreboard or styrene trays, one layer deep, holding 6 kg of fruit. Precooling assists in preserving fruit quality.

Processing characteristics

Avocados do not readily lend themselves to processing though success has been achieved in some areas. Peeled halves can be sealed in polythene with a nitrogen atmosphere and frozen, and a product called 'guacomol' is available in California which is essentially a savoury dip.

Oil is extracted from reject fruit and used in the cosmetic industry.

References

Anon. (1976), 'Avocado varieties for commercial planting in California'. Co-operative Extension Services, University of California.

Bergh, B.O. (1961), Breeding avocados at C.R.C. California Avocado Association Yearbook 45, p. 67.

Bergh, B.O. (1969), Avocado, *in* 'Outline of perennial crop breeding in the tropics' F.P. Zerweda and F. Wit (eds), Landbouwhogesch, Wageningen, p. 23.

Blumenfeld, A., and Gazit, S. (1974), Develoment of seeded and seedless avocado fruits. *Journal of the American Society for Horticultural Science* **99,** 422.

Chandler, W.H. (1958), 'Evergreen orchards'. 2nd edn, Lea & Febiger, Philadelphia, chapter 7, p. 205.

Chaplin, G.R. (1977), Storage procedures. Proceedings Australian Avocado Research Workshop, p. 107

Clarke, O.I. and Clarke, A.B. (1926), Results of pollination and other experiments on avocado at Point Lima Homestead, California. Avocado Association Report 1925–1926, 85.

Goodall, G.E., Embleton, T.W., and Platt, R.G. (1965), Avocado fertilization. Rev. Calif. Agr. Ext. Serv. Leaflet 24.

Gustafson, D. (1976), 1976 world avocado production. California Avocado Association Yearbook 60, p. 74.

Hodgson, R.W. (1947), Bearing habits of the avocado. California Avocado Association Yearbook 31, p. 35.

Kosiyachinda, S., and Young, R.E. (1976), Chilling sensitivity of avocado fruit at diferent stages of the respiratory climacteric. *Journal of the American Society for Horticultural Science* **101,** 665.

Lahav, E. and Kalmar, D. (1977), Water requirements of avocado in Israel. II. Influence on yield, fruit growth and oil content. *Australian Journal of Agricultural Research* **28,** 869.

Overby, J. (1978), Clonal rootstocks — modern science nips root rot in the bud. Avocado Grower p. 26.

Pegg, K.G. (1976), Root rot control in tropical rainforest soils. California Avocado Association Yearbook 60, p. 188.

Petersen, P.A. (1956). Flowering types in the avocado with relation to fruit production. California Avocado Association Yearbook 40, p. 174.

Platt, R.G., Goodall, G.E. and Lee, B.W. (1970), Thinning avocado orchards. University of California Agricultural Extension Bulletin AXT-n3.

Popenoe, W. (1920), The avocado, *in* 'Manual of tropical and sub-tropical fruits'. Hafner Press, New York, p. 9.

Sedgley, M. (1977), The effect of temperature on floral behaviour, pollen tube growth and fruit set in the avocado. *Journal of Horticultural Science* **52,** 135.

Whiley, A.W. and Brown, B.I. (1980), Avocado maturity times in Queensland. *Queensland Fruit and Vegetable News* **51,** 242.

Zerner, E. and Gottreich, M. (1975), Observations on the fertilization process in avocado with fluorescent light. *Euphytica* **24,** 531.

Chapter 13

LECYTHIDACEAE

BRAZIL NUT by D. McE. ALEXANDER
Bertholletia excelsa Humb. & Bonpl.

PARADISE NUT
Lecythis zabucayo Aubl.

Botany

This family consists of a group of large evergreen or deciduous trees of which two genera, *Lecythis* and *Bertholletia,* produce fruits which are used in commerce.

Bertholletia is represented by *B. excelsa* which is a rainforest tree with a very tall undivided trunk 30 to 50 m high. It has a rounded canopy standing above the remainder of the forest but sometimes occurs in pure stands. Its habitat extends over a vast area including parts of Guiana and Venezuela around both northerly and southerly tributaries of the Amazon in Brazil and into Bolivia. It has alternate, short stalked, unifoliate leathery leaves (25 to 30 cm x 10 to 15 cm). The flowers occur in upright panicles. They have 2 sepals, 6 equal pale yellow petals, numerous stamens which are united on a thick flap on the lower side of the flower and a short style. The fruits are globose and from 12 to 15 cm in diameter. They have a thick woody outer shell with 12 to 24 triangular nuts arranged inside like segments in an orange. Each nut is enclosed by a thick woody shell and a thin brown testa and consists of an oily endosperm which is rich in protein (14%) and vitamin B.

Several species of *Lecythis* namely *L. zabucayo* Aubl., *L. usitata, L. ollaria, L. pisonis* Camb., *L. urnigera* Mart, *L. lanceolata* Poir and *L. elliptica* Kunth produce edible nuts.

L. zabucayo is the paradise nut, sapucaia nut or monkey pot of commerce although sometimes *L. usitata* and *L. ollaria* are also included.

L. zabucayo Aubl., is a large forest tree up to 50 m high, from the Guianas and Brazil. It has elliptical unifoliate leaves about 15 cm long which are acuminate and are shed towards the end of the dry season. Flowers are white and are formed on terminal racemes. Unlike Brazil nuts its fruits which are urn-shaped have a thick woody lid which is shed when the nuts are mature and allows single nuts to fall to the ground. This occurs when the slender fleshy funiculus which supports each nut shrivels and decays on exposure to the atmosphere. The nuts are about 5 cm x 2.5 cm with thin brown shells which are less sharply angled than Brazil nuts and have longitudinal ridges. The kernel is ivory white and almost creamy in texture. It is sweeter than that of the Brazil nut.

Unlike the Brazil nut which fruits only on its topmost branches, the paradise nut crops throughout its canopy. According to Kennard and Winters (1960), *L. elliptica* is a highly ornamental small tree which often commences fruiting when only 2 m tall. Its capsules contain about 8 seeds and the nuts are chestnut brown marked with yellowish anastomosing lines.

Climatic requirements

These species thrive in an equatorial climate with a mean annual temperature of about 25°C, with little seasonal or diurnal temperature fluctuation and with a high and continuous rainfall totalling 1700 to 2700 mm per annum. In some parts of its natural habitat there is a short dry season of 1 to 2 months.

Soil requirements

Both species will grow on a range of soils including elevated areas but they grow best in deep, moist alluvial soils near watercourses. They do not thrive in situations subjected to regular or extensive flooding.

Genetic diversity

Considerable differences in yield, nut size and fruit size occur both between trees and between locations. According to Howes (1948), 17-year-old trees in Malaysia produced from 2 to 200 fruits.

Production season and time of cropping

In Brazil, nut collection commences in November and extends into June. The Acre territory which extends to the Bolivian border is an area of early maturity. The peak of nut collection is from January to March. The trees flower over a long period and nuts take 14 to 18 months to mature.

Yield potential

Young trees may take from 7 to 10 years to produce their first fruit while good mature trees yield up to 300 fruits, that is, 225 to 450 kg of nuts.

Market acceptability

Very good. In common with all nut crops there is an unsatisfied world-wide demand. Nuts have high oil ($> 60\%$), protein and vitamin B contents.

Export potential

Excellent, although competition for price from collections from wild trees could keep prices down.

Cultural details

Propagation

Propagation is by seed. No grafting or selection of desired clones is practised. In the wild, seed is spread by two species of *Agouti,* a squirrel-like animal which buries nuts as a food source. Forgotten nuts often germinate to produce new trees.

Management

Malaysian experience suggests a spacing of about 12 m between trees. Trees have been established at Singapore, Malayan peninsula, Sri Lanka and Trinidad.

Harvesting methods

Brazil nut fruits do not open naturally and are collected and opened with an axe or a machete. Nuts are dried, then washed and re-dried in preparation for marketing. Paradise nuts are collected in competition with wild animals (monkeys, wild pigs) and spread on mats and allowed to dry thoroughly.

Marketing characteristics

Nuts are marketed either in shell or as kernels. Most of those imported 'in shell' are shelled prior to sale.

Storage characteristics

It is important to keep nuts, both in shell and as kernels, cool and dry to prevent moulds developing.

Transport characteristics

Nuts are transported in bags in well-aired holds.

Processing characteristics

Shelling is done between rollers after nuts are graded for size.

References

Howes, F.N. (1948), 'Nuts — their production and everyday uses'. Faber and Faber, London.
Kennard, W.C. and Winter, H.F. (1960), Some fruits and nuts for the tropics. USDA Ag. Res. Serv. Misc. Pub. 801.

20 Avocado — Hazzard

21 Avocado — Sharwil

22 Avocado — Hass

23 Avocado — Fuerte

24 Sapucaia Nut

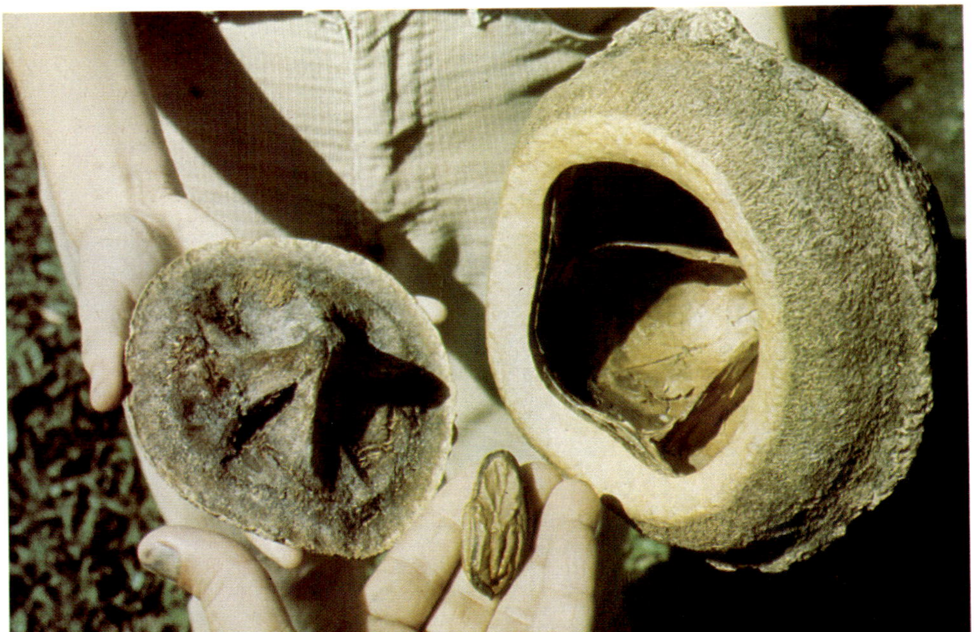

25 Sapucaia Nut

26 Duku

27 Langsat

28 Santol

29 Santol

30 Chempedak

31 Jackfruit

32 Breadfruit

33 Marang

34 Jaboticaba

35 Wax Jambu

36 Guisaro

37 Brazil Cherry

38 Guava

39 Feijoa

Chapter 14

LEGUMINOSAE

INGA

Inga spp.

by K.R. CHAPMAN

Botany

The *Inga* spp. are grouped under the subfamily Mimosoideae of the Leguminosae family. The genus appears to consist of 30 to 40 species of trees from the Guianas, Surinam, the West Indies, central America and Peru. They grow to at least 30 m high. The species are of importance as trees for wood and shade and a few members produce edible fruits namely *Inga edulis* (the ice cream bean) and *Inga laurina* (Sw.) Willd. (the sackysac).

Inga spp. have collectively been called chimbillo in Peru and chalum in the central Americas.

I. edulis is grown as a shade tree for coffee in Brazil and many were lost in the 1968 drought which hit Brazil.

The *Inga* spp. are known to fix nitrogen and *I. jinicuil* in experiments in Mexico was shown to produce root nodules and to be more effective at fixing nitrogen than *I. leptolober,* when used as shade trees for coffee.

I. punctata was investigated in 1978 for a source of anticancer agents.

I. chorrerona was collected in the Amazon River area of the Vitoto tribe in 1976 as a fruit-bearing cultivated plant.

I. laurina (Sw.) Willd., the sackysac, is a native of the West Indies and central America having edible fruits which are often sold in local markets. Botanically, it has pinnate leaves, usually with 2 pairs of leaflets and small white, fragrant flowers in spikes 10 to 12 cm long. The pods are oblong flat 5 to 15 x 2 to 3.5 cm with sweet white, slightly aromatic pulp, which is eaten (Purseglove 1969).

I. paterno is another species listed, which has edible fruits.

I. spectabilis and *I. acuminata* are market crops in Venezuela.

Climatic requirements

No specific requirements are listed in the literature, but the conclusion is that the Inga is a tropical rainforest tree susceptible to drought. It presumably prefers warm conditions and would probably succumb to frost. In Honduras it grows at 0 to 2000 m elevation.

Soil requirements

Again, no information is available. We presume that it grows in soils that drain well and are high in organic matter and probably acidic in nature.

Genetic diversity

No named varieties are reported in the literature and the diversity is unknown.

Production season and time of cropping

Many species produce year round; recognized cultivars mainly produce in summer.

Yield potential

Yield potential is unknown, but possibly is as good as tamarind at 150 to 200 kg fruit per year per tree.

Market acceptability

Apparently the fruit is quite acceptable to indigenous communities — at least for *I. edulis* and *I. laurina.*

Cultural details

I assume that these would be similar to those for tamarind, although, again, no detailed information is available.

Pests and diseases

Anastrepha distincta Greene (a fruit fly) is known to attack *I. feuillei* (the paceae) in Peru.

Automolis inexpectata (Roths) a lepidoptera has caused more damage to *I. feuillei* (the paceae) in Peru.

Cuscuta reflexa Roxg. is a known parasite of *Inga* spp.

Harvesting methods

The fruit is harvested by hand.

Marketing characteristics

The fruit would probably be marketed as fresh fruit pods.

Storage and transport characteristics

No information is available on storage and transport characteristics but they would probably be similar to tamarind.

References

Purseglove, J.W. (1969), 'Tropical crops — Dicotyledons Vol. 1', Longmans. Green and Co. Ltd., London p. 208.

Further reading

Anon. (1980), *Inga* spp. Annotated Bibliography (1931–1980). *Qld. Department of Primary Industries Bibliography Series* **80–41,** pp. 8, refs 25.

Bressam, R., Elias, L.G., Molina, M.R., Navarrete, D. (1977), Composition and potential use of some tropical fruits. *Archivos Latinamericanos de nutricion* **27,** 475.

Graner, E.A., and Godoy Jr. C., (1971), Shading for coffee plantations. V. Results of two more two-year cycles. 1967-68 and 1970-71. *Anais da escola superior de agricultura "Luiz de queiroz"* **28,** 153.

Roskoski, J.P. (1978), The importance of nitrogen fixation from non-crop legumes in a tropical agro-ecosystems. *Bull. Ecol. Soc. of America* **59,** 52.

Schultes, R.E. (1976), Plants of Colombia. XIX new fruit bearing cultivated plants from the Amazon River area of the Witoto Tribe. *Macoubea witotorum, Inga chorrerana. Bot. Mus. Leafl. Harv. Univ.* **24,** 193.

TAMARIND

by K.R. CHAPMAN

Tamarindus indica L.
Tamarindus occidentalis Gaertn.
T. officinalis Hook

Botany

Tamarinds belong to the subfamily Caesalpinioideae. The tamarind *(Tamarindus indica* L.) is native to the drier savannah of tropical Africa and possibly southern Asia. It grows wild throughout the Sudan and was long ago introduced into India. The fruit was well known to the ancient Egyptians and the Greeks in the 4th century B.C. It is now widely spread throughout the East Indies, Pacific, tropical America and West Indies.

The tamarind tree, a handsome, massive, evergreen tropical legume, ascends, under favourable conditions, to a height of 24 m and may attain a spread of 12 m and a trunk circumference of 7 m. It is slow growing and wind resistant, with strong, supple branches, gracefully drooping at the ends, and has dark grey, rough fissured bark. The mass of bright green, fine feathery foliage is composed of pinnate leaves, 8 to 16 cm long, each having 10 to 20 pairs of oblong leaflets 1 to 2.5 cm long and 5 to 6 mm wide, which fold at night. Inconspicuous, 2.5 cm wide flowers, borne in small racemes, are 5 petaled (2 reduced to bristles), yellow with orange or red streaks. The flower buds are distinctly pink due to the outer colour of the 4 sepals which are shed when the flower opens.

The fruits, flattish, bean-like, irregularly curved pods, are borne in great abundance and usually vary from 5 to 18 cm in length and from 2 to 3 cm in diameter. Exceptionally large tamarinds are found in Abaco, Bahamas. The pods may be cinnamon brown or greyish brown externally and, at first, are tender-skinned with green, highly acid flesh and soft, whitish, underdeveloped seeds. As they mature, the pods fill out somewhat and the acidulous pulp turns brown or reddish brown. Thereafter, the skin becomes a brittle, easily cracked shell and the pulp dehydrates naturally to a sticky paste enclosed by a few coarse strands of fibre extending lengthwise from the stem. The fully formed seeds are hard, glossy brown, squarish, 10 to 12 mm in diameter, and each is enclosed in a parchment-like membrane.

The flowers are thought to be pollinated by insects. The pulp which is about 55% of the fruit has a composition of: water 20.6%; carbohydrate 70.8% (30 to 41% reducing sugars); protein 3.1%; fibre 3.0%; fat 0.4%; ash 2.1%.

The acidity of the pulp is due to tartaric acid, 10%. Ripe pulp is rich in Ca, P, vitamin B and riboflavin and is a good source of niacin and thiamine.

The shell and fibre account for 11.1% of the fruit and the seeds about 33.9%. Tamarind seeds contain 63% starch, 14 to 18% albuminoids and 4.5 to 6.5% of semi-drying oil.

The tree is long lived and in Africa and India some trees have been estimated to be more than 200 years old.

Climatic requirements

Tamarinds grow throughout the tropics and subtropics as shade trees and fruit trees along roadsides, in parks and backyards.

It will withstand salt spray, and low temperatures of around –1 to –2°C, and more when an adult tree. Older trees are more cold tolerant than mango, avocado, and lychee. When small it is very susceptible to frost. The tamarind can withstand drought better than frost.

Dry weather is apparently important during fruit development, and in south Malaysia, when frequent rains occur at this time, the tamarind apparently does not bear. However, sufficient rain or irrigation is required in drier areas to produce good growth.

It is reputed to be insufficiently hardy to be grown in California.

Soil requirements

The tamarind thrives best in deep rich alluvial soils, but still does well even in rocky soils with little or no cultural attention.

Genetic diversity

Although some authors note various varieties of tamarind in India (which appears to be the only country with major plantations of tamarind) most are still apparently propagated by seed. It is not known if plants come true to type from seed, from various varieties or selections. However, it is suggested that wide variability appears to exist among seedlings.

Thailand has some named varieties, namely, Muen Chong, Nazi Zad, Sri Chompoo. Some tamarinds are reported as being much sweeter than others.

Production season and time of cropping

Tamarinds produce new leaves in spring, flower throughout the summer and pods appear in winter and ripen in early spring. Seedling trees are slow to come into bearing and take up to 10 to 12 years to produce an appreciable crop.

Yield potential

Mature trees in India produce about 150 to 200 kg of fruit per year. Fully grown trees reach 30 m in height and have a girth of up to 10 m.

Production for India, which apparently is the chief regular commercial producer of fruit was estimated at 250 000 tonnes in 1964.

Market acceptability

The tamarind has an exceedingly wide range of uses ranging from food and beverages to medicinal and industrial uses, and as a timber. Leaves, flowers, fruits, bark and roots are all used for various processed products. However, many of these uses are of doubtful economic importance, for example, production of ethanol, tartrates and pectin from the pulp.

While many medicinal claims are made for various preparations from fruit, leaves, flowers, bark, roots etc., only a few are apparently well recognized, for example, the antiscorbutic properties of the pulp; laxative action of the fruit juice; diuretic properties of the leaf sap; refrigerants for fevers and carminatives.

Fruit is often sold unshelled as it has good keeping qualities. Alternatively, at harvest the pulp is cleaned from the shell and seeds and fibre are removed and this cleaned pulp is compressed and packed in palm leaf mats for storage and marketing.

More important is the use of tamarind pulp as a food and beverage and the use of seeds to produce sizing powders for use in the jute and textile industries. These sizing powders are only half as costly as starch.

For shipment to processors the pulp is removed, layered with sugar and then covered with boiling sugar syrup. In east India the pulp is covered with salt, rolled into balls, sun dried, exposed to dew and stored in stoneware

Export potential

Australian export potential is probably limited. Most existing export is as cleaned pulp from India. In 1964 India exported 3000 tonnes; 50% went to Arab countries and the rest of Sri Lanka (formerly Ceylon), Europe and the U.S.A.

Morton (1958) reported that annual imports by the U.S.A. for medicinal use, were around 100 tonnes.

Reddish fleshed pulp is apparently preferred in Oriental markets for preserving.

Cultural details

Propagation

Propagation is primarily by seed. Seeds remain viable for many months and germinate within a week of planting. However, tamarinds can be propagated by cuttings, shield budding, side veneer grafting and air layering.

Management

Young plants should be protected from frost. It is usual to field plant seed, and protect the young seedling. Plants need watering during their first two summers for survival and weeds should be kept under control.

No publications appear to exist on irrigation and fertilizing of tamarinds, or any other aspects relating to training, fumigation, mulching, spacing etc. (Annon. 1980).

Pests

A number of insect pests are known to attack tamarinds. These include:

- *Lochmaeocles* spp. (Cerambycidae) which attack limbs.
- *Kerria lacca* (Kerr), a lac insect which attacks stems, twigs and leaves.
- *Calandra linearis* var. *striata* Thung, a beetle that attacks the fruit.
- *Charaxes fabius* Fabr., a butterfly larva that attacks leaves.
- *Caryedon serratus* (Oliver), the ground nut bruchid or tamarind weevil, which attacks pods, seeds and pulp in the field and roasted seeds in storage.
- *Taragama siva* (Lef.), a larva of a lepidopterous insect that attacks leaves of tamarind.

Nematode pests which attack tamarind are:

- *Radopholus similis*, a burrowing nematode.
- *Meloidogyne incognita*, rootknot nematode.

Diseases

Bartalinia robillardoides Tassi causes a leaf spot disease. *Aspergillus flavus* Link causes yellow mould disease of tamarind.

Bark extracts of tamarind are known to suppress the infectivity of three strains of watermelon mosaic virus namely VB, M and C.

Fruits of tamarind have been shown to possess antimicrobial activity — antibacterial and antifungal.

Harvesting methods

Pods or fruits are cinnamon or greyish brown externally and are tender skinned with green highly acid flesh with soft white underdeveloped seeds when immature.

At maturity the pods fill and the pulp turns brown to reddish brown. The skin becomes brittle and the shell is easily cracked. The pulp dehydrates and becomes sticky, and the seeds become hard and glossy brown, enclosed in their parchment-like membrane. Fruits are ready to harvest in early spring and are hand harvested from the trees.

Marketing characteristics (see also market acceptability)

The fruit is sold either as fresh fruit or as pulp in the fresh market place. Both have good keeping qualities.

Storage characteristics (see also market acceptability)

Data on length of storage are not available, but must be at least 3 to 6 months or more as pulp, covered with syrup to allow shipping to far off destinations.

Transport characteristics

Tamarinds can be transported both as fruit or pulp over long distances.

Processing characteristics (see also market acceptability)

With respect to food products the following items are processed from tamarinds:

- Ingredients in chutneys, curries, sauces, pickles.
- Sugared confectionery of tamarind pulp.
- Patties and sweet meat.
- Tamarind ade, shakes, syrups, shaved ice confectionery, sherbet and ice cream
- Canned jelly and jam, nectar, and fruit preserves in combination with guava, papaya and banana.

References
Anon. (1980), Tamarind *(Tamarindus indica)* — Annotated Bibliography (1931–1980). *Qld. Dept. of Primary Industries Bibliography Series* **80–38,** pp. 14, refs 79.
Morton, Julia F. (1958), The tamarind *(Tamarindus indica* L.*)*. Its food, medicinal and industrial uses. *Proceedings of the Florida State Horticultural Society* **71,** 288.
Further reading
Fisch, B.E. (1974), The tamarind. California Rare Fruit Growers Yearbook 1974, p. 221.
Leferve, J.C. (1971), Revue de la litterature sur le tamarinier. *Fruits* **26,** 687.
Lewis, Y.S., Dwarkanath, C.T.., and Johar, D.S. (1954), Utilization of tamarind pulp. *J. Sci. Indust. Res. New Delhi* **13A,** 284
Lewis, Y.S. and Neelakanton, S. (1964), The chemistry, biochemistry and technology of tamarind. *J. Sci. Indust. Res. New Delhi* **23,** 104.

Chapter 15

MALPIGHIACEAE

ACEROLA (BARBADOS CHERRY, WEST INDIAN CHERRY, PUERTO RICAN CHERRY, JAMAICAN CHERRY)

by K.R. CHAPMAN

Malpighia glabra L.
Malpighia punicifolia L; *M. emarginata* Moc. & Sesse ex DC.

Botany

The acerola *(Malpighia glabra* L.*)* or Barbados cherry as it is more commonly known in Australia, belongs to the *Malpighia* genus of the Malpighiaceae family. It is the best known fruiting member of the genus *Malpighia* which contains 30 to 40 species. The genus is indigenous to the area between southern Texas, U.S.A. through Mexico and Central America to Surinam, Peru and the West Indies.

Relatives of *M. glabra* of interest are the caapi. These are lianas of the Amazon basin which belong to the *Banisteriopsis* spp. and are a source of narcotics for the Indians. *M. mexicana* (A. Juss) or *M. edulis* (D. Smith) produces a large edible fruit, also called 'acerola' in Costa Rica. *M. suberosa* L. has been used as a nematode resistant rootstock for *M. glabra* L. with variable results.

The acerola is a dense, spreading, glabrous shrub 2 to 5 m high, with numerous branches. Branches are thick and woody and have conspicuous raised lenticels. Stipules are minute and deciduous with leaves opposite, shiny, dark green above, ovate to elliptic and 2.5 to 7.5 cm long, with short petioles. Leaves and stems become glabrous when mature. Flowers are 1 to 2 cm across, arranged in axillary cymes, hermaphrodite, calyx with 6 to 10 sessile glands, petals 5, fringed slender claw-like, pink or red, stamens 10.

The fruit is a bright red berry-like juicy drupe, depressed ovoid, obscurely 3-lobed, 1 to 3 cm in diameter, borne singly or in pairs or triplets, on short pedicels in leaf axils. Fruits have three seeds which are triangular and ridged. The skin is very thin and easily bruised. Average fruit weight is 7 to 10 g and is said to resemble the northern crab-apple in taste, due to the presence of malic acid. Sweetness varies from tart to sweet depending on clone, and has 60 to 70% extractable juice.

The fruit is very high in ascorbic acid (vitamin C) and varies from 1000 to 4676 mg (of ascorbic acid) per 100 g. It is only exceeded in ascorbic acid content by rose-hip *(Rosa rugosa)* which has between 1700 and 6977 mg per 100 g. The *Acerola* is a good source of vitamin A and Fe, a fair source of Ca and P and has small amounts of B vitamins, with total sugars between 3.5 and 4.7%. One fruit of *Acerola* will satisfy the daily requirements of an adult for vitamin C.

Climatic requirements

The acerola is susceptible to frosts, and about 1600 mm of rain per annum is optimal in Puerto Rico. High rainfall increases fruit yields but makes the fruit softer. The tree appears to be capable of growing throughout tropical and subtropical regions, from about 30°N to 30°S latitude. Flowering does not appear to be day-length sensitive.

Soil requirements

Acerolas will grow over a wide range of soil types from alkali rockland soils of Dade County to the acid sands of central Florida. However, a pH of greater than 5.4 is preferred in Puerto Rico where low pH soils give symptoms of calcium deficiency. Yield increases of up to 400% have been recorded in Puerto Rico where soils with a pH of 5.4 were limed with one tonne of lime per hectare. They grow well in Marl soils and in the clay soils of Puerto Rico.

Acerolas will not tolerate waterlogging. However, heavier soils are preferred because of susceptibility to nematodes, namely, *Meloidogyne incognita* (Kofed and White) var. *acrita* Chitwood, and *Radopholus similis* (Cobb) Thorne.

Genetic diversity

The acerola shows considerable variation when grown from seed, and selection from seedling populations has been used to produce superior clonal varieties in Florida. Ledin (1956, 1958) reports on clonal varieties selected in Florida and Puerto Rico, and makes mention of the Puerto Rican B-17, a tart variety, and Florida sweet, an agreeable semi-sweet variety with apple-like flavour.

Desirable selection characters are: high yields which increase yearly; large fruit with high juice content; thick skin and firm flesh to reduce bruising; desirable flavour, tart or sweet, for processing or fresh fruit respectively; erect open growing trees; easily propagated from cuttings (Ledin 1956).

Puerto Rican B-15 surpasses B-17 for both productivity and ascorbic acid level (Ostendorf 1963).

With acerola the seed is only formed as a result of normal fertilization, but in many cases seed is non-viable.

Hybridization with *M. mexicana* may be useful for increasing fruit size, and with *M. suberosa* L. for nematode resistant rootstocks.

Production season and time of cropping

Fruiting season extends over 3 to 7 months, depending on clone, commencing in mid-spring and extending to late autumn. Plants flower continuously with peaks of heavy flowering and heavy cropping. These peaks vary from 3 to 4 to 8 to 10, depending on clone.

Vegetatively propagated plants will flower and fruit the second year in the field, but will not bear heavily until the third or fourth year after planting. Fruits mature 3 to 4 weeks after flowering.

Yield potential

Puerto Rico B-17 produces 9 to 13 tonnes per hectare at 4 years. Florida sweet produces around 8 tonnes per hectare at 5 years and up to 57 tonnes at 8 years from 300 plants.

Puerto Ricon B-15 produces greater yields than B-17 but the author does not have data on these yields.

Market acceptability

As a fresh fruit the acerola is very tender and perishable and should be used as soon as possible after harvest.

It is widely used as a fresh fruit, particularly in the West Indies and Central and South America. In the Caribbean the acerola is known as the 'tree of life' and is regarded as having exceptional nutritive and tonic properties.

It can be processed into fresh juice, frozen ice, purees, juice blends, baby food syrups, flavouring syrups, ice cream and preserves, and vacuum dried powders containing 30 to 38% w/w of ascorbic acid. Pasteurized canned juices do not keep well, are unattractive, and the CO_2 produced swells cans and ascorbic acid is quickly lost. Frozen pieces are far more acceptable and retain ascorbic acid levels. The fruit is not very satisfactory for pies and conserves as the stones do not separate readily from the flesh.

Puerto Rico had an industry based on plantations of over 400 hectares in the early 1960s to produce raw material for nutritional use. This was mainly in the form of dehydrated powder to be used later, diluted to juice, and sucrose-free syrups for children as a source of ascorbic acid. The industry failed in 1967 as agricultural labourers moved to the cities, and despite investigation of mechanisation of harvesting the industry did not revive. Charley (1970) concluded that on economic grounds the use of synthetic vitamin C is more advantageous and therefore, in the search for new fruits for fruit production, flavour is more important. On the other hand, yields of 500 to 2000 kg per hectare could be realized.

Export potential

As fresh fruit it would be very limited because of the highly perishable nature of the fruit. As processed products, acerolas would have some export potential as a source of ascorbic acid, but of course, have to compete with synthetics.

Cultural details

Propagation

Air layers: In spring or summer as for lychees.
Cuttings: Mist propagation of leafy hardwood cuttings 12 to 30 cm long and 0.5 to 1 cm in diameter. Use IBA at 5000 ppm as basal dip.
Grafting: Side veneer or cleft graft are rarely used as techniques.
Seeds: Stones are cleaned of flesh, allowed to dry, dusted with a seed protectant for damping off and planted in an uncracked state.

Management

Fumigation: Where light soils are used and nematodes are present, fumigate prior to planting. Follow-up treatments with nematicides may be necessary with established plants.
Planting: At 6 to 12 months old, just before rainy season.
Spacing: 4 to 5 m x 2 to 5 m in rows for hedging or distinct plants. Can be interplanted with avocado, mango, lychee, etc.
Mulching: Heavy mulching recommended and weed control is required.
Pruning: Plants producing thick, dense growth should be thinned to promote yield. Upright growers (for example Florida sweet) have open growth and these should be headed back to promote side branching and to promote yield. Prune in early autumn after fruiting and apply some fertilizer after pruning to promote winter plant health. Spring pruning reduces yields.
Fruit set: 50 ppm PCPA induces good fruit set and partial parthenocarpy.

Irrigation

Irrigation is essential for maximum growth and yields of large fruit. Spring irrigation in dry months is particularly important for cropping. Irrigation is not necessary in the late autumn and winter, when plants are dormant.

Fertilization

Liming is mentioned in soil requirements. Early spring fertilizing encourages good growth and flowering. Apply about 0.25 kg of 10:0:10 per tree per year of age to year 6 in August.

Additional fertilizer is applied in October, November and January using a 4:7.5:3 mixture or 6:4.6:3 for older plants at a rate of 0.5 kg per tree per year of age per application. For older trees above 10 years apply 2.5 kg of 10:0:10 in early spring and 5 kg for each application in October, November and January.

Zinc sprays of a neutral zinc compound may be required, where general yellowing and slow-down of growth occurs with deficiency.

Cibes and Samuels (1955) describe mineral deficiency symptoms of acerola.

Pests and diseases

Soft brown scale (*Coccus hespridium* Linn.), green scale (*C. viridis* [Green]), cottony cushion scale (*Icerya purchasi* Mask.) and two other soft scales (*Saissetia* sp. and *Pulvinaria urbicola* Ckll.) attack acerola.

Plant bugs (*Nezara viridula* L. and *Leptoglossus phyllopus* L.) damage fruits.

The butterfly larva (*Ephyriades brunnea* var. *floridensis* B and C) feed on the leaves of acerola.

As mentioned earlier in soil requirements, nematodes are a pest of acerola and are by far the most serious pest. *M. suberosa* L. is resistant to root knot nematods, but grafts of acerola on this stock have produced dwarfed, unthrifty trees. Hybrids between *M. suberosa* and acerola may prove more useful.

Use of heavier soils and fumigation are necessary for acerola.

Harvesting methods

The harvest season extends over 3 to 7 months and fruit must be picked every second day.

The fruit is ready for harvest when it turns pink or red. Completely ripe fruits spoil quickly after harvest and cannot be transported.

Hand harvesting is the only method used at present and because of the very thin skin the fruit is easily damaged. Some damage to fruit is tolerated if it is quickly processed. However, mechanical harvesting is difficult with present varieties, as the fruits mature at different stages over an extended period.

Marketing characteristics

The fruit is attractive when fully ripe but very susceptible to damage. The sweeter varieties may appeal for the fresh fruit market.

Poor transportability will limit fresh fruit market exposure. With regard to processed products, the range is impressive and the attraction is the vitamin C content.

Storage characteristics

Half ripe fruit will hold for several days under refrigeration.

Frozen juices or pieces of fruit appear to keep for extended periods. Vacuum dehydrated concentrate powder will keep for over 12 months in cans under refrigeration when a desiccant bag is added.

Pasteurized canned juice is a poor product with a short storage life and is not recommended.

Transport characteristics

Transport characteristics are generally poor in the fresh fruit state.

References

Charley, V.L.S. (1970), Some aspects of modern fruit juice industry. *Voedingsmiddelen Technologie* **1,** 421.

Cibes, H. and Samuels, G. (1955), Mineral deficiency symptoms displayed by acerola trees grown in the greenhouse under controlled conditions. Puerto Rico University A.E.S. Tech. Bull. 15, 1.

Ledin, R.B. (1956), A comparison of three clones of Barbados cherry and the importance of improved selections for commercial plantings. *Proceedings of the Florida State Horticultural Society* **69,** 293.

Ledin, R.B. (1958), The Barbados or West Indian cherry. Univ. Florida Agr. Exp. Sta. Gainesville, Bull 594,1.

Ostendorf, F.W. (1963), The West Indian cherry. *Tropical Abstracts* **18,** 145.

Further reading

Anon. (1980), *Malpighia* spp. Annotated Bibliography (1969-1980). *Qld. Dept. Primary Industries Bibliography Series* **80-37,** pp. 18, refs 49.

Charley, V.L.S. (1969), Some tropical fruit juices. Tropical Products Institute Conference Papers, p. 161.

Stahl, A., Kaplow, M., and Nelson, R. (1955), The present status and future possibilities of Barbados cherries. *Proceedings of the Florida State Horticultural Society* **68,** 138.

Chapter 16

MELIACEAE

SANTOL by B.J. WATSON
Sandoricum koetjape (Burm. f.) Merrill
Sandoricum indicum Cav.

Botany

Santol is a well-known Asian fruit of the family Meliaceae and is lately described as *Sandoricum koetjape* (Burm. f.) Merrill. In older texts it is usually described as *Sandoricum indicum* Cav. A red Malaysian form (kechapi santol) was previously described as *S. nervosum* but is also now included under *S. koetjape*. The 10 species in the genus *Sandoricum* have origins in Mauritius, India, Burma, Thailand, Kampuchea, Malaysia, Indonesia, Borneo and the Philippines. *S. koetjape* is native to Malaysia, Indonesia, Borneo and the Philippines. Additional common names are kechapi and sentul (Indonesia, Singapore and southern Malaya), setul and sentoi (northern Malaya) and ka-thon (Thailand).

Seedling trees are quick growing, reaching 30 m in height with a smooth straight trunk. Asexually propagated trees tend to be smaller (to 20 m) with a more bushy habit of growth. Leaves are dark green, compound, with 3 leaflets, each up to 200 mm long. In the dry season period just prior to flowering the tree becomes semi-deciduous, and mature leaves turn yellow or red before dropping.

Flowers are small, up to 12 mm wide in loose hanging bunches arising from ancillary buds along the branches. The flowers are lightly scented, the sepals form a small cup and there are 5 separate, reflexed petals and 10 stamens, which are joined into a tube with the anthers at the top, the stigma just showing above them. Fruits are round, slightly flattened, 50 to 100 mm across with a thick firm and downy rind and slight lengthwise indentations. The external rind colour varies from dull to golden yellow, sometimes with a tinge of pink. The aril is relatively thin (up to 15 mm) but white, juicy and transluscent, and surrounding and attached to 3 to 5 seeds about 15 to 20 mm long.

The two basic types of santol are, according to Molesworth Allen (1965), red and yellow. Yellow santol is distinguished by the fact that leaves turn yellow before falling and fruit pulp is generally sweet. Red santol has red leaves at maturity, longer leaflets (up to 300 mm) and sour aril. Most selected cultivars appear to have arisen from the yellow santol form.

Climatic requirements

The tree is relatively cold tolerant and is grown at up to 1200 m altitude in Java (Sastrapradja 1977) which is higher than tolerated by trees such as soursop *(Annona muricata)* which is grown at up to 1000 m altitude. Thus it should be able to be grown in frost-free areas as far south as southern Queensand. Campbell *et al.* (1977) rated damage on mature trees in southern Florida following a severe frost of −3°C as very severe, but damaged trees regenerated even when all but the below ground portion was killed. The santol is grown and crops well in southern Florida according to Whitman (1974).

The prime environment for best clones appears to be in a hot, wet monsoonal climate with a dry season in winter and early spring.

The tree can survive and crop relatively well in areas with rainfall down to 800 mm per annum.

Soil requirements

The santol does not appear to require a high level of soil fertility. However, well-drained soils are preferred. If soil pH levels exceed 7 the tree can suffer severe iron deficiency

Genetic diversity

There is considerable variation in tree form, fruit size and aril quality amongst the various wild forms and also in seedlings from selected clones. In the Philippines the small 'native' variety, which was selected from a wide gene pool, has now been surpassed since introduction of Thai clones (Asis, 1971). Thailand appears to have been the leader in clonal selection and whilst there is still not a substantial amount of aril in the best selections, they are a vast improvement over selections in other countries. The 'Manila' santol referred to by Florida writers is in fact a Thai cultivar (probably Khan Tong or Teparod) now commonly known in the Philippines as 'Bangkok'.

The Philippines 'native' variety is reputed to have a genetic chromosome number of 11 whereas 'Bangkok' has 22.

Production season and time of cropping

In latitudes greater than 8° north or south of the equator, the santol flowers in the spring and matures in summer. Florida harvest is in August and September, and in the Philippines and Thailand, from July to October. In Malaysia the season is earlier — in June and July. Northern Australia (above 18°S) could expect maturity in January to March and in more southerly areas the season could be later. The few bearing trees in Cairns flower in August-September and fruit mature in February, the development period being 5 to 6 months. Asexually propagated trees commence bearing in the third to fifth year after field planting.

Yield potential

No yield data are available. A few clonal trees are established in north Queensland but the majority are still juvenile.

Market acceptability

Fruit from good santol cultivars have a flavour which appeals to most people at first taste. The fruit is also thirst quenching. However, even with the best clones the aril recovery is relatively poor and there is difficulty in separating this from the seeds. Thus the prospects for the fruit other than a backyard tree or for tropical fruit 'tourist packages' are not likely to be great.

Export potential

The fruit is unlikely to have potential for export.

Cultural details

Propagation

Cultivars are propagated by air layers; approach, wedge, whip, or side veneer grafts and by T or modified forkert budding. All methods give good results.

Mature trees can be top worked with methods commonly used for deciduous fruits and this is best attempted during the 'dormant' period in late winter to early spring.

Management
Trees seldom if ever require pruning. Wind-breaks in exposed areas are a distinct advantage since santol branches are easily broken in strong winds.

Irrigation
Irrigation is probably necessary for maximising yield in very dry (less than 1000 mm per annum) zones, but in areas with regular summer rainfall distribution irrigation would not be as critical as with most other tropical tree crops.

Fertilization
There is no documented information, but organic mulches would be beneficial for conserving moisture and providing nutrients, particularly around young trees.

Pests and diseases
Very little information on pests and diseases is reported. Whitman (1974) states that in Florida the main pest is the Caribbean fruit fly *Anastrepha suspensa.* However, although the fly oviposits on ripening fruit causing an unsightly fly speckled appearance, the thick rind prevents the larvae developing.

In north Queensland the fruit spotting bug *(Amblypelta lutescens)* damages stems of flowers and young fruit.

Harvesting methods

Fruit are most commonly harvested on the basis of external colour and softness, or picked following the commencement of natural drop.

Marketing characteristics

The tough thick rind prevents major damage during handling of fruit. However, this obviously leads to some abuse in Asia and fruit are usually marketed in a 'scuffed' condition. The fruit are also usually relatively cheap and this encourages bulk handling.

Under Asian ambient summer temperatures the fruit can be held in the market for up to 7 days without serious deterioration.

Storage characteristics

Data on storage under refrigeration are not available.

Transport characteristics

As explained previously the thick fruit rind ensures little damage from bulk handling.

Processing characteristics

Little information on processing is available. The aril is preserved by bottling in syrup in the Philippines and also makes a good jam. However, extracting the aril requires considerable hand labour. The rind is sometimes candied and is a good product. There is no commercial canning as such.

References

Asis, C.V. (1971), 'Plants of the Philippines'. Science Education Centre, University of the Philippines Press, Quezon City.

Campbell, C.W., Knight, R.J., and Zareski, N.L. (1977), Freeze damage to tropical fruits in southern Florida in 1977. *Proceedings of the Florida State Horticultural Society* **90,** 254.

Molesworth Allen, B. (1965), 'Malayan fruits'. Eastern University Press Ltd, Singapore.

Sastrapradja, S. (1977), 'Buah-buahan'. Lembaga Biologi Nasional, Bogor.

Whitman, W.F.C. (1974), The camu camu, the 'wan' maprang and the 'Manilla' santol. *Proceedings of the Florida State Horticultural Society* **87,** 375.

LANGSAT and DUKU
Lansium domesticum Jack

by B.J. WATSON

Botany

The genus *Lansium,* which belongs in the family Meliaceae contains 6 or 7 species native to India, Burma, Malaysia, Indonesia, Borneo and the Philippines. Langsat and duku are classified as a single species, *Lansium domesticum* Jack, but there are differences in tree form, fruit arrangement on the raceme, and the fruit itself. Another form, reputed to be a hybrid, is called duku-langsat, but appears to lack a precise definition.

Langsat and duku are thought to originate in Malaya and possibly also in Sumatra and Borneo. Common names are langsat and longkhong (Thailand); langsat, dukuh and duku (Indonesia and Malaysia); lanzon, lansones and lanzones (Philippines); and lanson (Caribbean).

Langsat is a slender tree, 10 to 20 m high with a straight trunk, slender upright branches and an irregular or rounded crown. Duku tends to be less tall but has a wider crown. Leaves of langsat are large, and divided into 5 to 7 alternate leaflets, which are dark green and dull below and measure 150 to 200 mm long by 60 to 100 mm wide. Leaves are faintly hairy underneath (langsat) or hairless (duku). Flowers are borne in racemes, 2 to 5 of these racemes grouped together directly on the trunk or large branches. These cauliflorous racemes are erect at first but later droop after fruit set and during fruit development. Racemes are 100 to 300 mm long, being smaller on duku than langsat. Individual flowers are approximately 12 mm wide and 5 mm long, and number 20 to 30 per raceme. Flowers are perfect and subsessile. There are 5 sepals and 5 petals, and the stamens are united to form a fleshy tube with 10 anthers. Flowers have a sweet smell. In all studies conducted, pollen has been found sterile.

Fruit are slightly ovoid in langsat and round in duku, and 30 to 50 mm and 40 to 50 mm in diameter respectively. There are commonly 15 to 25 fruits per raceme in langsat and 4 to 12 in duku. The developing fruit is pale green, ripening to a greyish buff, pale muddy yellow of pale gold but frequently with brown blemishes. The pericarp is thin in langsat and contains a sticky latex. Duku has a thicker pericarp (up to 6 mm) and no latex.

The skin of both fruit peels easily and cleanly from the clear, white, translucent, and juicy aril. In langsat flavour varies from sweet to sour but duku is generally very delicate and sweet. The duku fruit consists of five separate segments, up to two of which (usually the larger) contain soft green bitter seeds. Langsat has from 1 to 5 seeds which are firmly attached to the aril. According to Bernardo *et al.* (1961), pollen grains of langsat are sterile and fruit and seeds are respectively autonomously parthenocarpic and apomictic in development.

Climatic requirements

The main Asian production areas are within 15° of the equator. Almeyda and Martin (1977) state that the crop cannot be grown on the main Asian continent except in southern Thailand and Vietnam. However, Uttaradit in the northern central plains of Thailand (17°N) is in fact a notable centre for langsat production. Also, langsat has been fruited in Florida (24°N) by Whitman (1979).

Langsat is recognized to be more cold tolerant than duku. In Java, both langsat and duku are cultivated to an altitude of 650 m (Soeseno 1968; Anon. 1975). At 600 m in Java, and 300 m in southern Luzon (Philippines) mean minimum yearly temperature is approximately 19°C and yearly normal is 22°C. This in broad terms corresponds with a latitude of 16 to 20°S in coastal Queensland, although the latter would have a number of nights with much lower temperatures (down to 9°C) than at altitude in the tropics. However, providing temperatures do not fall near to 6°C at any time, growth (as evidenced at Kamerunga Horticultural Research Station) should be satisfactory as far south as 20° in Queensland. Langsat and duku are suspected to be damaged with temperatures below 6°C. However, temperatures of up to 40°C are not detrimental.

The langsat and more particularly duku have moderately high soil water requirement, although langsat can tolerate some dry spells. However, in the absence of adequate rainfall irrigation is necessary to prevent flower and fruit drop.

Soil requirements

Langsat and duku can grow successfully in many soil types, but for best performance a well-drained sandy or clay loam rich in organic matter is preferred (Coronel 1977).

Genetic diversity

An appreciation of diversity is complicated by the uncertainty of whether one is dealing with one species or several subspecies or forms. Also, seed is formed apomicticly in cultivated forms and it is assumed that wild types must have had pollen transfer or were quite heterozygous in order to give the present variations in form (Bernardo *et al.* 1961). Cultivar selection is practised in Thailand, Indonesia, Malaysia and the Philippines and each country has a small number recommended. Selection has been based on fruit flavour and sweetness, low percentage seed development, total yield per tree and spread of harvest season, Paete, a langsat type in the Phillipines, is very highly regarded, as is uttaradit (langsat) and longkhong (duku) in Thailand.

Characteristics of the main Philippine selections are shown in table 1.

Table 1. Lanzones characteristics in the Philippines

Langsat (l) and Duku (d) Cultivars —	Fruit length mm	Fruit width mm	Fruit weight g	T.S.S. %	Number seed per fruit	Tree habit
Paete (l)	32	27	15	16.5	0 to 2	Upright
Mindanao (l)	35	32	32	13.2	1 to 2	Spreading
Duku (d)	35	30	30	17.8	0 to 1	Spreading

(Coronel 1977)

In Malaysia and Indonesia there is a wider diversity of forms of langsat, langsat-duku, duku and kokosan (the latter a small fruited, large seeded form).

Production season and time of cropping

Langsat and duku in the more northern latitudes of the tropics (greater than 10°N) normally have one crop a year, maturing in the period April to October. The few trees producing in Florida ripen fruit between August and October. A dry period in spring appears to stimulate differentiation and flowering occurs at or shortly after the commencement of the monsoonal rains.

In Malaysia there are 2 crops a year, but often only one when a dry period extends past the normal period of flowering. In Java (south of the equator) the crop matures in the period November to February, during the summer monsoon. Hawaiian trees (few in number) mature in autumn to early winter. By interpolation, maturity in northern Australia would probably be between February and May.

The period of development from flowering to fruit maturity appears to be from 3 to 4$^1/_2$ months, varying according to climatic conditions.

Yield potential

The long juvenile period of trees and low production are serious limiting factors to development of the crop in Australia

Grafted trees are stated by Coronel (1977) to take from 5 to 9 years to commence bearing, but large air layers only 2 to 4 years. Trees from seed commence bearing at 10 to 15 years. Yields are most often quoted as 5 kg in the ninth year from planting, and up to 30 kg per tree for 25 to 30-year-old trees. Spacing recommendations vary from 8 to 12 m. However, Coronel suggests 5 to 6 m (330 trees per hectare) which at 25 years could produce 8.25 t per hectare. Average yields in Thailand in 1975 (106 000 tonnes from 11 560 ha) are quoted as 9.2 t per hectare (Vangnai personal communication 1979). The average number of fruit produced on a mature tree in various countries ranges from 500 to 1500.

Market acceptability

In its favour, the langsat, duku or duku-langsat has several redeeming characteristics: fruit are easy to peel; the aril is soft and moist but not extremely juicy or sticky, it is free of fibre and often with minimal number of seeds.

In good cultivar selections, there is an agreeable balance between acidity and sweetness — unusual in many tropical fruits. However, many fruits marketed in Asia are bitter, and the skins affected by anthracnose. European reaction to the fruit varies from favourable to poor, but good selections are usually acceptable.

Export potential

There would only be export potential from Australia if yields, fruit quality and post-harvest life were all excellent. This proposition is unlikely.

Cultural details

Propagation

There are no major problems with propagation other than that seedlings may require at least 1 year of growth before grafting is possible. Langsat seedlings are generally preferred as rootstocks. Approach, cleft, whip or side veneer grafts commonly give from 60 to 90% success. Air layers (large branches preferred) are usually made at the beginning of the wet season and detached 5 to 6 months later. Softwood or semi-mature terminal cuttings will strike under mist with a fair success rate. Both I.B.A. and N.A.A. are beneficial.

Management

A humid microclimate is essential for growth and for this reason wind-breaks are mandatory in northern Australia. In addition, shade is essential in the first 2 to 3 years after planting, although only overhead and westerly protection is necessary. *Leucaena leucocephyla* appears to be a satisfactory shade tree at Kamerunga Research Station.

Trees are seldom pruned in Asia apart from removing crossing or dead and damaged limbs. However, heading back langsat trees 2 to 3 times during the first 3 years provides a more branched structure.

Irrigation

Supplementary watering is necessary in northern Australia, and would be critical after flowering since in stressed trees developing fruit either drop or atrophy on the branch. Langsat and duku have a relatively short lateral root distribution (2.4 m from the trunk) but grow to a depth of 1.2 m.

Fertilization

Langsat and duku are slow growing trees and to maximise growth for early bearing regular but limited fertilizing is suggested. Precise fertilizer recommendations are not available but NPK mixtures high in N and P up to first flowering and high in N and K subsequently, applied after fruit set and again after harvest each year, are suggested by Coronel (1977). Mulching with bagasse appears beneficial in north Queensland.

Pests and diseases

The larvae of the green moth (*Prasinozena* spp.) and the carpenter moth (*Cossus* spp.) infest and damage the flower buds in the Philippines. Anthracnose *(Colletotrichum gloeosporioides)* is a pandemic problem on fruit, both pre- and post-harvest.

In north Queensland main damage is from defoliation caused by larvae of various lepidoptera species. The most serious pathogen is collar rot at ground level caused by *Stilbella cinnabarina.* Control of the latter is with a copper oxychloride slurry application.

Harvesting methods

Fruit maturity is indicated by the change from green to yellow pericarp, but the actual picking stage colour is best determined by experience. All fruit on the tree do not mature together, but generally all fruit on one raceme do. Racemes are removed with secateurs or a knife. Individual fruits are then separated or packed for market intact on the raceme.

Marketing characteristics

Fruit require care in handling at all stages since (particularly with langsat) bruising adds to skin discoloration. Grading for size is desirable. Generally all fruit are eaten fresh but there is a small crystallizing industry in the Philippines.

Superior cultivars (for example, paete in the Philippines) are generally recognized by the buying public in Asia and prices vary considerably according to the cultivar and time of the year. Few details of nutritional value are available, but there is negligible vitamin C content in the fruit.

Storage characteristics

Anthracnose *(Colletotrichum gloeosporioides), Rhizopus nigricans* and *Aspergillus luchuensis* all contribute to post-harvest rots. However, *Aspergillus* can only infect injured fruits (Srivastava and Mathur 1955).

Sound fruit remains attractive for a maximum of 4 days under ambient (25 to 33°C) temperatures. Skin browning is the main problem. However, Pantastico *et al.* (1969) found that fruit would store substantially free of rots and browning for a period of up to 2 weeks if treated with a cold benomyl dip (4000 ppm ac) and held at 15°C in an atmosphere of 5% O_2, nil CO_2 and 85 to 90% relative humidity.

Transport characteristics

Fruit can be likened to the banana in handling properties and generally in Asia containers are generally too large and not rigid enough to prevent damage. A loose fill pack in a rigid container not exceeding 200 mm depth would appear to be satisfactory.

Processing characteristics

There are no reports of any substantial existing or prospective processing industries.

References

Anon. (1975), 'Bertanam Pohon Buah-Buahan 2'. Penerbitan Yaysan Kanisius, Yogyakarta, 2nd edn.

Almeyda, N. and Martin, F.W. (1977), Cultivation of neglected tropical fruits with promise. Part 4, The Lanson. Agr. Res. Service, U.S.D.A.

Bernardo, F.A., Jesena, C.C. and Ramirez, D.A. (1961), Parthenocarpy and apomixis in *Lansium domesticum*. *Philippine Agriculturist* **44,** 415.

Coronel, R.C. (1977), Growing of lanzones. Dept. Hort. College Ayr. Univ. Philipp. Los Banos, Ext. Circ. 14.

Pantastico, E.B.., Mendoza, D.B. and Abilay, R.M. (1969), Some chemical and physiological changes during storage of lanzones (*Lansium domesticum* Correa). *Philippine Agriculturist* **52,** 505.

Soeseno, S. (1968), 'Buah-Buah Dikebun Rumah'. P.T. Kinta, Djakarta

Srivastava, H.C. and Mathur, P.B. (1955), Studies in the cold storage of langsats, *Journal of the Science of Food and Agriculture* **6,** 511.

Whitman, W.F.C. (1979), Rare Fruit Council International Inc. Yearbook, Museum of science and Planetarium, Miami, Florida.

Chapter 17

MORACEAE

by M. SEDGLEY

Artocarpus is a genus of about 50 species, most of which are native to Asia and at least 15 of which have edible fruit. The fruit provides a staple starchy food in many tropical parts of the world and trees have been cultivated since antiquity. The most important species is *A. altilis*, the breadfruit, but there is evidence of introgression between this species and *A. mariannensis* and there is probably sexual compatibility with other species also. There is also graft compatibility within the genus.

Breadfruit is consumed mainly as a vegetable and will be considered separately from jackfruit, chempedak and marang which are eaten as fruits. The Amazon tree grape will also be described separately although very little information is available on this plant.

BREADFRUIT, BREADNUT
Artocarpus altilis (Parkins.) Fosb.
Artocarpus incisus L.f.

Botany

The tree which grows up to 20 m high, is evergreen in the everwet tropics but deciduous under prolonged dry conditions. It is monoecious with male and female inflorescences on specialized laterals. The species is outbreeding and all parts of the tree contain latex.

The leaves are dark green, entire towards the base and 3- to 9-lobed towards the tip. They are 30 to 60 cm long. The individual male flowers are minute with a single stamen. Large numbers are produced on the dense, yellow, club-shaped, pendulous catkins which are about 25 cm long. The individual female flowers have a tubular calyx and a 2-celled ovary. The flowers are grouped together into a large upright head on a spongy receptacle. The resulting fruit is a syncarp consisting of the fused ovaries of the numerous female flowers. It is 10 to 30 cm in diameter, green, yellow or brown with a hexagonally marked rind. The fruit may be seeded or seedless. The seedless fruit is the more common. It is grown for the flesh which is cooked as a vegetable. The cause of the seedlessness is unknown but is probably a result of long cultivation. The seeded form is often called the breadnut. It has little edible flesh but the seeds are cooked and eaten. The seeds are flattened and about 2.5 cm long.

Climatic requirements

The breadfruit is a species from hot, humid, tropical lowlands. It is very cold sensitive and will not grow if the temperature falls to 5°C. It prefers a regular rainfall of between 150 and 250 cm per year.

Soil requirements

The breadfruit will grow on a variety of soils provided they are of sufficient depth and are well drained. Best growth occurs on deep moist soils with a high humus content and high fertility.

Genetic diversity

There is high variability within the species in characters such as fruit quality, time to start of bearing, keeping quality of fruit and bearing season. The breadfruit has been profoundly modified during centuries of cultivation. There is evidence of introgression with *A. mariannensis* (Fosberg 1960) and triploidy is common amongst the seedless clones (Barrau 1976). However, there has never been any plant breeding or clonal selection effort although a collection of Oceanic breadfruit clones was established for the purpose 10 years ago in Western Samoa. Clones are numerous but local in distribution. There is probably some sexual compatibility between the various *Artocarpus* species. Breadnut seedlings show high variability.

Production season and time of cropping

Fruits are produced throughout the year. The peak season in India, Hawaii and the West Indies is during June to September with a small crop in winter. The clonal breadfruit tree starts to bear at around 6 years old. The seedling breadnut bears after 8 to 10 years.

Yield potential

The tree produces regular crops. A 10-year-old tree will produce 500 to 700 fruits per year each weighing 2 to 5 kg. Thus the yield potential is 1000 to 3500 kg of fruit per tree per year.

Market acceptability

Some forms become sweet on ripening and are eaten as a fresh fruit but the breadfruit is mostly eaten ripe or unripe as a vegetable. It can be boiled, roasted, steamed, fried, baked or made into soups or dough. It is mainly a carbohydrate source consisting of 70 to 75% water, 20 to 30% carbohydrate, 1.5% protein and 0.5% fat. It contains vitamins A, B and C but not in large amounts (Oliviera 1974).

As the breadfruit is large and used mainly as a vegetable, its market acceptability in Australia would probably be low. The seeds of the breadnut can be boiled or roasted and are said to resemble chestnuts. They could have more consumer appeal than the fruit.

Export potential

There is demand in some western countries such as the United Kingdom but there are problems in exporting the fresh product due to the extreme perishability of the fruit (Bell and Coursey 1971). Breadfruit slices in brine are canned in Jamaica and shipped to the U.K. The fresh or preserved product could probably be exported to Asia from Australia.

Cultural details

Propagation

The traditional method or propagation of seedless forms is by root suckers lifted from around an established tree. Root cuttings are now more commonly used, as plants can be produced more quickly (Lopez and Rodriguez 1975). Stem cuttings can be induced to root using IBA. Ground and air layering are also possible. Grafting and budding are not generally used although the breadfruit can be approach grafted onto the same species or onto *A. heterophyllus* or *A. elasticus*. It can be budded on *A. rigidus* and *A. elasticus*, and also on *A. altilis* (Rowe-Dutton 1976).

The breadnut is propagated by seeds which lose viability quickly and should be sown immediately.

Management
A spacing of 10 to 15 m is recommended. The trees are fast growing and live for 30 to 40 years. They have moderate wind and salt tolerance, poor drought and flood tolerance and no cold tolerance. Young plants grow better under shade but older plants require full sun. Vegetatively propagated plants produce a surface mass of feeding roots which sucker easily. The suckers should be suppressed under cultivation. The breadfruit is probably wind pollinated but insect pollinators may improve set as hand pollination in India has proved very successful.

Irrigation
Regular irrigation is needed during drought.

Pests and diseases
Very few pests and diseases have been reported despite the long history of cultivation. Soft rot of the fruit caused by *Rhizopus artocarpi* has been reported in India and termites and brown scale have attacked the trees in Darwin (Hearne 1975). A disease known as pingelap has appeared in the last 20 years in the islands of Micronesia. It is characterized by either a sudden foliage wilt or by dieback. The casual agent is unknown (Barrau 1976).

Harvesting methods

The fruit is not usually collected until it has fallen to the ground. This does not appear to damage the fruit or shorten its storage life.

Marketing characteristics

The fruit has a very short shelf life of only a few days. This would lead to problems in marketing the fresh fruit.

Storage characteristics

The fruit is highly perishable and is stored traditionally by submerging in water. This method increases the storage life by only a few weeks and causes splitting of the fruit. Other traditional storage methods include fermentation, sun drying and artificial drying. The fruit is stored cooked in Polynesia. Storage in polyethylene bags at temperatures around 12°C extends the storage life by about 2 weeks. Chilling injury occurs below 12°C (Thompson, Been and Perkins 1974).

Transport characteristics

The high perishability and poor storage characteristics indicate that transport of the fresh fruit would be limited to a few weeks at the most under ideal conditions.

Processing characteristics

Breadfruit slices in brine are canned in Jamaica in small quantities to supply local and overseas markets. Breadfruit can also be dried (Barrau 1956) and a Californian study found that dehydration of the sliced or diced fruit by tunnel drying or freeze drying was feasible (Reeve 1974). The reconstituted fruit could be used as a vegetable. A further possibility is the use of the fruit as a source of commercial starch.

The seeds of the breadnut can be roasted or boiled and thus may be amenable to processing.

References
Barrau, J. (1956), Drier for preservation of breadfruit. *SPC Quarterly Bulletin* **6,** 16.
Barrau, J. (1976), Breadfruit and relatives, *in* 'Evolution of crop plants', N.W. Simmonds (ed.), Longman, London.
Bell, J.M.K. and Coursey, D.G. (1971), Tropical vegetable in Britain. *Tropical Science* **13,** 251.
Fosberg, F.R. (1960), Introgression in *Artocarpus* (Moraceae) in Micronesia. *Brittonia* **12,** 101.
Hearne, D.A. (1975), 'Trees for Darwin and northern Australia'. Watson Ferguson and Co., Brisbane.
Lopez, S.J. and Rodriguez, S.J. (1975), A method to obtain relatively uniform breadfruit trees from a stockplant. *Univ. Rio Pedra J. Ag.* **59,** 77.
Oliviera, J.F.S. (1974), The nutritional value of some foods consumed on San Tome Island. *Ecology Fd. & Nut.* **3,** 237.
Reeve, R.M. (1974), Histological structure and commercial dehydration potential of breadfruit. *Economic Botany* **28,** 82.
Rowe-Dutton, P. (1976), *Artocarpus altilis* — breadfruit, *in* 'The propagation of tropical fruit trees'. R.J. Garner and S.A. Chaudri (eds), Commonwealth Ag. Bur.
Thompson, A.K., Been, B.O., and Perkins, C. (1974), Storage of fresh breadfruit. *Tropical Agriculture* **51,** 407.
Further reading
Purseglove, J.W. (1974), 'Tropical crops — Dicotyledons'. Longman, London.

JACKFRUIT
Artocarpus heterophyllus Lam.
Artocarpus integer (Thumb.) Merrill; *A. integrifolius* of Bailey not L.f.

CHEMPEDAK
Artocarpus integer (Thumb.) Merrill

MARANG
Artocarpus odoratissima Blanco.

Botany

All species are monoecious evergreen latex-producing trees which grow up to 20 m high. The dark green leaves are entire in jackfruit and chempedak and lobed in marang. The chempedak is distinguished from the jackfruit by the smaller fruit and by long wiry brown hairs on the leaves and twigs. The two may be different forms of the same species. The individual male and female flowers are small and grouped together in inflorescences. In the jackfruit the female inflorescences arise from the trunk and older branches. The fruits are syncarps. That of the jackfruit is the largest of all cultivated fruits reaching 90 cm in length. It is elongated with green, yellow or brown rind with short sharp hexagonal spines. There are 100 to 500 seeds per fruit and the pulp surrounding the seeds is strong smelling in the jackfruit and chempedak. The marang fruit is round to oval and 15 to 17 cm in diameter. The outer rind consists of soft fleshy yellow-green spines. Inside the flesh has the appearance of white grapes where it surrounds the seeds. There are no seedless forms of any of the species.

Climatic requirements

All species are tropical and require a warm humid climate. The jackfruit is the most cold tolerant species as it can withstand some frost. Both the chempedak and the marang are severely damaged by temperatures below 5°C.

Soil requirements

The jackfruit and chempedak will grow on a variety of soils provided they are well drained. Best production occurs on deep alluvial soils of open texture and pH 6.0 to 7.5. There is no information available for the marang.

Some chempedak cultivars

Prepared by M. Sedgley

Cultivar Name/ Number	Origin	Yield	Fruit characteristics	Comments
CH19	Selangor	Moderate	Large, long and curved. Rind yellow-green. Pulp thick and yellow with coarse texture and strong aroma.	
CH27	Selangor	High	Large, long and straight. Rind yellow-green. Pulp thick, yellow and sweet.	
CH28	Selangor	High	Medium-large. Rind yellow. Pulp thick, yellow-orange, sweet with good flavour, fine texture and small seed.	
CH29	Selangor		Medium-large and straight. Rind green. Pulp thick, orange and sweet with fine texture.	Highly attractive fruit due to orange flesh.

Translated by Dr. R. Winoto Suatmadji from a paper by Mohd. Nor bin Ibrahim, Mohamed bin A. Shukor and Lye Tuck Thye.

Some jackfruit cultivars

Prepared by M. Sedgley

Cultivar Name/Number	Origin	Fruit characteristics	Comments
J2	Malaysia	Large. Pulp yellow-green, sweet and slightly acid with coarse hard texture and little aroma. Fruit always splits.	Poor storage qualities
J27	Malaysia	Medium-large. Pulp orange, fine textured and sweet.	
J28	Malaysia	Medium-large. Pulp yellow, thick and sweet.	
J29	Malaysia	Medium-large and globular. Pulp yellow, thick, sweet and firm, fruit low in latex.	Good for fresh consumption
J30	Malaysia	Large. Pulp yellow, thin, sweet and slightly acid with fine texture.	
J31	Malaysia	Small and elongated. Pulp yellow and sweet, with fine texture and strong aroma. Fruit is high in latex.	Suitable for canning

Translated by Dr. R. Winoto Suatmadji from a paper by Mohd. Nor bin Ibrahim, Mohamed bin A. Shukor and Lye Tuck Thye.

Genetic diversity

Both the jackfruit and chempedak show high variability in characters such as time to seed germination, fruit shape, colour, flavour, texture, and amount of latex, time to start of bearing and time of maturity of the fruit. There has been no clonal selection or plant breeding effort and there are numerous local types. The most notable are a small fruited Indian type called Rudrakshi and Singapore jack which bears after 18 months from sowing and ripens later than other types. There is no information available on the marang.

Production season and time of cropping

Jackfruit and chempedak start to bear between 3 and 8 years after sowing. The marang bears after 6 years. Fruit is produced all year with the main crop in summer in all species.

Yield potential

Jackfruit trees produce up to 250 fruits per year each weighing 10 to 30 kg although 50 kg fruits have been recorded. Thus, a single tree will produce up to 2500 to 7500 kg of fruit per year. However, jackfruit and chempedak have a tendency to biennial bearing. There is no information available for the marang.

Market acceptability

All three species can be eaten ripe as a fruit or unripe as a vegetable. The seeds of all species can be eaten boiled or roasted. The edible pulp of jackfruit and chempedak comprises 30% of the fruit and consists of 73% water, 0.6% fat, 23% carbohydrate, 2% fibre and 0.5% ash. The seeds comprise 5% of the fruit and consist of 52% water, 7% protein, 0.5% fat, 40% carbohydrate, 1.5% fibre and 1.5% ash.

The market acceptability of jackfruit and chempedak as a fresh fruit would probably be low in Australia due to the large size and strong smell of over-ripe fruit. The marang would be more acceptable due to its smaller size and reportedly sweet, juicy, aromatic and pleasant flavoured flesh. Consumer education would be needed before the unripe fruit could gain favour as a vegetable. The seeds of all species would probably be acceptable in the processed form.

Export potential

There is no information available. However, the jackfruit and chempedak are familiar in Asia and there would probably be a demand for the fresh product.

Cultural details

Propagation

At present all species are largely propagated by seed which loses its viability quickly and must be planted immediately. Soaking for 24 hrs in water, 25 ppm NAA or 100 to 500 ppm GA improves and hastens germination. These species do not commonly produce root suckers, as does the breadfruit, although root cuttings can be used to propagate all species.

Stem cuttings and air layers are both very successful in jackfruit provided 5000 ppm IBA is used on etiolated tissue (Mukherjee and Chatterjee 1979). Jackfruit can be budded onto *A. integer* and *A. rigidus* as well as onto its own species. It can be grafted or inarched onto *A. integer, A. hirsuta* and *A. altilis* as well as onto *A. heterophyllus*. However, there are reports of graft incompatibility with *A. altilis* and *A. hirsuta*. Chempedak can be budded onto *A. heterophyllus, A. rigidus* as well as onto the same species. Approximately 12-month-old rootstocks are used for grafting and budding of both species (Rowe-Dutton 1976). There is no information on grafting or budding of marang.

Management

None of the species is currently grown in orchards but 6 to 12 m spacings are suggested for jackfruit and chempedak. The trees are slower growing than breadfruit, have moderate wind and salinity tolerance and poor cold, drought and flooding tolerance.

The seedlings are difficult to transplant as the tap root is easily damaged. Thus seedling rootstocks should be grafted in situ. The trees have a tendency to biennial bearing and probably require insect pollinators. There is no information available on the marang.

Irrigation

Regular irrigation is required during drought.

Fertilization

For jackfruit and chempedak the Malaysian recommendation is for N:P:K:Mg (8:4:2:1) at 30 g per tree at 6 months with a doubling of the quantity each 6 months up to 2 years. Thereafter 1 kg of N:P:K:Mg (4:2:4:1) for each tree every 6 months (Kanapathy 1976). There is no information available for marang.

Pests and diseases

No pests or diseases have been reported for marang and very few reported for chempedak. Jackfruit, on the other hand, is reportedly attacked by an enormous number of pests and diseases. These included shoot borers, bark borers, bud weevils, spittle bugs, mealy bugs, scale insects, aphids and fruit rot, seed rot, blossom rot, leaf spots and pink disease (Butani 1978). The main problem in Florida is fruit rot caused by *Rhizopus artocarpi* (McMillan 1974), and in Darwin aphids, scale, mealy bugs, grasshoppers and wood borers have been reported (Hearne 1975).

Harvesting methods

The fruit is allowed to fall to the ground before harvesting.

Storage characteristics

Jackfruit can be stored in the processed form. There is no information on storage of fresh fruit.

Processing characteristics

Jackfruit is canned in syrup in India and Indo-China and the ripe fruit can be used for jams, jellies and ice cream manufacture. The unripe fruit could probably be dried in a similar manner to breadfruit for use as a vegetable. The fruit can also be frozen. The seeds can be boiled or roasted.

References

Butani, D.K. (1978), Pests and diseases of jackfruit in India and their control. *Fruits* **33**, 351.

Hearne, D.A. (1975), 'Trees for Darwin and northern Australia'. Watson Ferguson and Co., Brisbane.

Kanapathy, K. (1976), Guide to fertilizer use in Peninsular Malaysia. Bull. 143, Min. of Ag. and Rural Devel. Malaysia.

McMillan, R.T. Jnr. (1974), *Rhizopus artocarpi* rot of jackfruit *(Artocarpus heterophyllus). Proceedings of the Florida State Horticultural Society 1974*, **87**, 392.

Mukherjee, S.K. and Chatterjee, B.K. (1979), Effects of forcing, etiolation and indole butyric acid on rooting of cuttings of *Artocarpus heterophyllus* Lam. *Scientia Horticulturae* **10**, 295.

Rowe-Dutton, P. (1976), *Artocarpus heterophyllus*, Jackfruit, *in* 'The propagation of tropical fruit trees'. R.J. Garner and S.A. Chaudri (eds), Commonwealth Ag. Bur.

Further reading

Purseglove, J.W. (1974), 'Tropical crops — Dicotyledons'. Longman, London.

California Rare Fruits Council Yearbook 1977.

AMAZON TREE GRAPE
Pourouma cecropiaefolia Mart.

Botany

Pourouma cecropiaefolia is a small dioecious evergreen tree native to Brazil with fruit produced in large racemes. The individual fruits are 4 cm in diameter with a thick red skin, sweet white flesh and a large seed.

Climatic requirements

The tree is native to the wet equatorial rainforests of South America and thus requires a warm humid climate.

Production season and time of cropping

The fruit ripens in autumn in Florida and over a 3 month period in the wet season in South America. Trees bear after 3 years.

Yield potential

The tree is reported to fruit heavily and be a prolific bearer.

Market acceptability

Market acceptability is probably high as the fresh fruit is attractive and sweet. However, the skin is acrid and inedible and must be removed before eating. It is reported to peel off easily.

Cultural details

Propagation
Propagation is by seed.

Management
Nothing is known of the agronomic requirements as it is cultivated singly around Indian houses in South America. It is fast growing, has moderate wind tolerance and poor cold and drought tolerance. The tree is dioecious so male and female trees should be interplanted for pollination.

Irrigation
Regular irrigation would be necessary during drought.

Pests and diseases
There are few if any pests and diseases in the native habitat.

Harvesting methods

The Indians use a forked branch to knock off the fruit.

Further reading
'Underexploited tropical plants with promising economic value'. National Academy of Science, Washington, 1975.

Chapter 18

MYRTACEAE

GENERAL ACCOUNT

by D.H. MAGGS

Botany

Apart from their fleshy fruits, the fruit-bearing Myrtaceae have much in common with our familiar eucalypts, although, because they have fewer xerophytic adaptations, the leaves are thinner, and in the American species, usually smaller and sometimes deciduous.

The flower is typically myrtaceous. The inferior ovary is closed by a disc through which the single stigma protrudes and around whose edge is the ring of stamens, 4 to 5 petals and the thick sepals. The ovary is 1 to 3 celled, with few to many ovules, and may be further subdivided into 2 to 6 cells. The seeds may lie free in the ovule or be embedded in a pulp.

Many of the fruit species occur in the large genus *Eugenia* and its split-offs *Syzygium* and *Myrciaria*. These all have 1 to 2 seeds in the ovary and may be distinguished from each other thus:

	Eugenia	*Syzygium*	*Myrciaria*
Petals and sepals	4	4–5	4
Calyx ring prolonged above the ovary disc	no	yes	yes
Sepals absciss	no	no	yes
Seeds free in ovary cavity	yes	no	no

Syzygium is an Indo-Malaysian genus and *Myrciaria* and most *Eugenia* species originate from America and the West Indies. The other two major fruit genera *Psidium* and *Feijoa* have compound ovaries containing many seeds, and *Psidium* has 5 petals and sepals while *Feijoa* has only 4. The leaves of *Feijoa* are white on the underside with dense short hairs. Other genera are *Ugni* with a single fruit species, *U. molinae*, in Chile, and *Rhodomyrtus* with *R tomentosus* from India.

Probably nearly 100 myrtaceous fruit species have been recorded of which about 50 are in the more generally available literature. Since their nomenclature is confused, information is presented in the following section on the most important species in the hope of establishing a consensus agreement on their common names and scientific equivalents. For this purpose, I have taken Bailey's Hortus Third as being a generally reliable and available authority, although it is by no means without errors and idiosyncrasies.

Most entries are restricted to nomenclature plus a few notes and references (cited by number), but more details is given for the most promising species. This choice was often difficult because one authority calls insipid a species which another strongly recommends. Such discrepancies, which may be found even in the earliest accounts, sometimes refer to climate or the condition (ripeness) of the fruit when tasted. It may be that in each would-be fruit area people will be obliged to make their own species evaluations.

Climatic requirements

The Indo-Pacific *Syzygium* species can be regarded as requiring warm tropical conditions with very low frost risk. The *Eugenia* and *Myrciaria* species will tolerate –2°C to–3°C, the *Psidium* species –5°C and *Feijoa* and *Rhodomyrtus* species –7°C.

Exceptions are *E. aggregata* (–7°C), *E. curranii* (0°C) and *P. friedrichsthalianum* (0°C). In dry, sunny areas a few degrees lower are often tolerable. Low temperatures may also inhibit full-sized fruits from developing their proper flavour and for those species such as guava which ripen part of their crop during May-June, monthly minima exceeding 10°C should be required during these months.

Soil requirements

Average horticultural soil is generally acceptable. *Rhodomyrtus* could be tried where drainage is poor but the soil reaction is not alkaline.

Genetic diversity

Superior local varieties have been reported for the more common species, but generally no effort appears to have been made to pick these out when the species has been introduced elsewhere. Polyembryonic seeds have been noted for several species. *P. guajava* has many varieties and there is considerable variation within progenies. (Basic x=22. Polyploid series reported for *Psidium, Eugenia.*).

Production season

Production season would probably be 2 to 4 months after the beginning of the wet season and may continue into early winter.

Yield potential

Most species appear capable of carrying heavy crops, sometimes 2 to 3 times a year. For commercial species yields between 10 and 30 tonnes per hectare are reasonable.

Market acceptability

Fruits are visually attractive and there should be few problems with the fresh fruit market. Most species are used for fruit drinks locally.

Export potential

Export potential is limited by storage life. There is ready acceptability in south-east Asian markets and a possibility of air-freight to luxury markets for best flavoured species.

Cultural details

Propagation
By either (a) Seed: — Polyembryonic varieties occur; (b) Cuttings: — semi-hard in mist, hardwood with bottom heat; (c) Grafting: — 2 to 3 year wood generally better than current. Rootstock sp. *Syzygium pycnanthum* is termite resistant.

Irrigation
Standard irrigation is probably necessary during growing season.

Pests and diseases
Myrciaria is said to be resistant to fruit fly because of its tough skin.

Harvesting methods

Cauliflorous and bushy species could be harvested by shakers. Hand picking is necessary for the taller species. Where flowering extends over several months fruit has to be picked twice a week for a long period.

Marketing characteristics

Thick-skinned species such as *Myrciaria* and some *Psidium* species are transportable. Others may be liable to bruising unless carefully packed.

Storage characteristics

Storage life can probably be measured in days.

Processing characteristics

Most species have been used for juices, purees and jams. (See also sections 2 and 3 on guava and feijoa.)

Bibliography

1. Allen, B.M. (1975), 'Common Malaysian fruits'. Longman, London.
2. Bailey, L.H. and Bailey, E.Z. (1976), 'Hortus Third'. Revised and expanded edition, MacMillan, N.Y.
3. Campbell, C.W. (1977), Cultivation of tropical fruits of the Myrtaceae in southern Florida. *Proc. Tropical Region ASHS* **21**, 37.
4. Cribb, A.B. and Cribb, J.W. (1974), 'Wild food in Australia'. Collins, Sydney.
5. Dijkman, M.J., Wehlburg, C., Ochse, J.J., and Soule, M.J. (1961), 'Tropical and sub-tropical agriculture'. MacMillan, N.Y.
6. Popenoe, W. (1934), Manual of tropical and sub-tropical fruits. Xerox copy, Univ. Microfilms, 1974.
7. Sturrock, D. (1959), Fruits for southern Florida. Stuart.
8. Uphof, J.C.T. (1968), 'Dictionary of economic plants'. Cramer, Lehre.

ENUMERATION AND NOMENCLATURE OF THE PRINCIPAL FRUCTIFEROUS SPECIES OF THE MYRTACEAE

1. **Britoa acida** Berg — PARA GUAVA
   Psidium araca hort. not Raddi in Florida, *P. acutangulum* (not *acidum*) DC., *Campomanesia acida* Berg.
   Tree 10 m, twig rectangular, leaves 5–10 cm, glabrous, flowers solitary, calyx opening irregularly. Fruit 5–7 cm, yellow, pulp acid, not musky, seeds few. Brazil 5, 6*
   Campomanesia spp., try *Britoa, Psidium.*
2. **Eugenia aggregata** (Vell) Kiaersk — RIO GRANDE CHERRY
   Myrciaria edulis hort. in U.S.A. but not *M. edulis* (Vell) Skeels.
   Brazil 2, 3
3. **Eugenia brasiliensis** Lam — GRUMICHAMA, BRAZIL CHERRY
   Eugenia dombeyi Skeels.
   Tree; seedlings fruit at 5 years old. Leaves 10 cm, flowers solitary on 5 cm pedicels; fruit 2 cm, red, green calyx, 1–2 seeds. The tree is frost hardy, easily propagated by seed or grafting. The fruits ripen in one month. Yields reported Philippines. This species is recommended by 3, 5, 6 on account of manageability and fruit flavour 2, 3, 5, 6, 7
4. **Eugenia curranii** C.B. Robinson — LIPOTE
   Syzygium currannii (Rob) Merrill. (*S. cumini* is the JAMBOLAN.)
   Tree 10 m, leaves 12–20 cm, flowers in short panicles, fruit 3–4 cm, purple, slightly acid. Considerable variation in fruit and habit. Sandy soils. Philippines 2, 3, 7
   Eugenia dombeyi Skeels, see *Eugenia brasiliensis* Lam.
   Eugenia edulis Benth, see *Myrciaria edulis.*
5. **Eugenia foliosa** D.C. — ARRAYAN
   Ecuador 8

* References numbered according to preceding bibliography.

6. **Eugenia Klotschiana** Berg. PERO DO CAMPO
Shrub, leaves, 7–10 cm, hard, brittle, silvery. Fruit pear-shaped, 5–10 cm long, slightly acid, aromatic. Brazil, savannah 6

7. **Eugenia ligustrina** (Swartz) Willd.
Tree shrub, dense foliage, leaves 5 cm, flowers solitary axillary in narrow bracts, fruit globose, 5 mm, red-black, sweet. South America, West Indies 2, 3

8. **Eugenia luschnathiana** Klotsch ex Berg. PITOMBA
Phyllocalyx luschnathiana (L) Berg.
Tree 10 m, leaves dark green, glossy, underside pale, flowers solitary, pedicel 2–3 cm, fruit globose, 2–3 cm, orange-yellow, acid, aromatic. Brazil 2, 3, 6

9. **Eugenia Selloi** Berg. PITANGA TUBA
Similar to *Eugenia uniflora*. Brazil 8

10. **Eugenia uniflora** L. BRAZIL CHERRY, SURINAM C., PITANGA
Eugenia michellii Lam, *Stenocalyx brasiliensis* Berg.
Tree to 7 m, bushy, leaves 3–5 cm, glossy, flowers 1 cm, axillary. Fruit flattened, 8 longitudinal grooves, 3–4 cm, red, aromatic, sweet, seedlings may be acid. This species is widespread throughout the tropics and subtropics. On alkaline soils it tends to be chlorotic. Often produces 2 crops per year. Low or zero setting may occur in isolated trees suggesting pollen incompatability mechanisms. Frost hardy. Fruits in 2–3 years. South America. 2, 3, 5, 6

11. **Eugenia uvalha** Cam UVALHA
Tree to 4 m, dense foliage, leaves oblong 3 cm. Fruit yellow, pear-shaped, 3–4 cm diameter, yellow, pulp watery, sweet 'onion' flavour. Brazil 2, 3, 6

12. **Feijoa sellowiana** Berg. FEIJOA, PINEAPPLE GUAVA
See following paper on feijoa by D. Batten.
GUAJAVA, GUAYAVA, see *Psidium guajava.*

13. **Myrciaria cauliflora** (DC) Berg. JABOTICABA
Eugenia cauliflora D.C.
Tree to 10 m, very slow growing, dense foliage, small thin leaves. Flowers in leafless tufts on older branches and trunk. Fruit globose, 2–3 cm, black, shining, pulp colourless, sweet; skin tough. Ripens in 20 days from flowering. Because of potential for heavy cropping and tough fruit skin, this species would be suitable for mechanical harvesting. Can be propagated by cuttings, grafting and seed. Frost retards but does not kill. Many varieties in Brazil. Closely related to *M. jaboticaba* and *M. paraensis*. Slow early growth could be limiting factor. South Brazil 2, 3, 5, 6, 7

14. **Myrciaria edulis** (Vell) Skeels.
Eugenia edulis Vell. For *M. edulis* hort. see *Eugenia aggregata.*
Tree 5 m, leaves willow-like. Flowers axillary/terminal clusters. Fruit 5 cm, orange 'oniony'. Brazil. . 2

15. **Myrciaria floribunda** (West ex Willd) Berg. GUAVA BERRY
Eugenia floribunda.
Tree 10 m, flowers in clusters, fruit globose, 1 cm, red-yellow, sweet, aromatic. Central and South America, West Indies 2, 3

16. **Myrciaria jaboticaba** Berg. JABOTICABA
Very similar to *M. cauliflora.* Flowers on long pedicels. South Brazil 3, 6, 8

17. **Myrciaria paraensis** Berg. CAMU CAMU
As *M. cauliflora.*
Flowers in leaf axils. Fruit 2–3 cm, dark red, acid. South Brazil 6

18. **Myrciaria vexator** McVaugh.
Similar to *M. cauliflora.*
Leaves dark green, flowers in leaf axils, fruit 3 cm globose, purple. Select for good flavour. Some seedlings said to be astringent. Brazil 3
Psidium araca Raddi, see *Psidium guineense* Swartz.
Psidium araca hort. in Florida, see *Britoa acida* Berg.
Psidium cattleianum, see *Psidium littorale* var. *longipes* and *Psidium littorale* var. *littorale.*

19. **Psidium friedrichsthalianum** (Berg) Niedenzu. COSTA RICAN GUAVA, CAS
Tree 8 m, shoots square, leaves glossy, 8 cm, fruit pear-shaped, 4–7 cm, yellow, white flesh, few seeds, slightly acid. Costa Rica 2, 3, 5

20. **Psidium guajava** L. GUAVA, APPLE GUAVA, PEAR GUAVA, COMMON GUAVA
P. pyriferum L. + *P. pommiferum* L.

21. **Psidium guineense** Swartz. GUISARO
Psidium araca Raddi, *P. molle* Bertol, *P. minus* Mart, *P. dichotomum* Weinmann. Note that *P. guineense* hort. Florida and *P. guinense* hort. California = *P. guajava* (Popenoe).
Shrub 3 m, foliage red pubescent when young, deciduous, flowers axillary 1–3 cm, fruit globose, 4 cm, yellow acid. Widely cultivated in tropical America and West Indies 2, 3, 6, 7, 8

22. **Psidium littorale** Raddi var. *littorale* Bailey YELLOW STRAWBERRY GUAVA, YELLOW CATTLEY GUAVA, WAIAWI

Psidium cattleianum Sab. var. *lucidum* (Degener) Fosb. *P. araca* hort. not Raddi, *P. acre* Ten., *P. chinense P. variable* Berg, vegetative characters as for *P. littorale* var. *longipes.*
Fruit globose, 2–3 cm, yellow, sweet, slightly aromatic. Fruits usually larger and less strongly flavoured than var. *longipes* q.v. 2, 3, 6

23. **Psidium littorale** Raddi var. *longpipes* (Berg) Fosb. STRAWBERRY GUAVA, PURPLE STRAWBERRY GUAVA

Psidium cattleianum Sab., *P. araca* hort. not Raddi, *P. variabile* Berg.
Bush 5 m, slow growing, leaves oval, lanceolate, smooth, 5 cm long. Flowers solitary. Fruit globose 1–2 cm, purple-brown, sweet, aromatic, seeds few, embedded in pulp. Ripening late summer before *P. guajava.* Usually heavy cropping. Propagated mainly by seed, graft, low percentage rooting from cuttings. Frost hardy. Fruit can be left several weeks on bush after ripening. Limited use for fresh fruit market. Widespread in tropics and subtropics 2, 3, 6, 7

24. **Psidium longipes** (Berg) McVaugh

Eugenia longipes Berg., *Mosiera bahamensis* (Kiaersk) Small, *Myrtus verrucosa* Berg.
Small shrub 1 m. Fruit black. (Included to prevent confusion with *Psidium littorale* var. *longipes.*) . . 2
Psidium lucidum hort. see *P. littorale* var. *littorale.*

25. **Rhodomyrtus tomentosus** (Ait) Hassk. DOWNY MYRTLE

Shrub 3 m. Flowers pink. Fruit globose 1–1.5 cm, green/purple, seeds few, pulp sweet, acid. Said to tolerate poorly drained soils. Very frost hardy (–7°C). Very ornamental. India. There are several Australian species of Rhodomyrtus 2, 3, 6
Jambosa, see *Syzygium.*
Stenocalyx, try *Eugenia.*

26. **Syzygium aqueum** (Burm) Alston WATER ROSEAPPLE

Jambosa aquea (Burm f.) DC.
Trees 10 m. Leaves glossy, dark green, 15 cm long. Flowers terminal/axil clusters 2–3 cm. White/red. Fruit pyriform 2–4 cm diameter, white/red, sweet, watery, tasteless, seeds 0–2. S. pycnanthum (termite resistant) may be used as a rootstock. Malaysia, Borneo, Cape York 2, 3, 5

27. **Syzygium cumini** (L) Skeels. JAMBOLAN, JAMBU, JAVA PLUM, LAMBOL

Syzygium jambolana (Lam) DC, *Eugenia cumini* (L) Druce, *Eugenia jambolana* Lam.
Tree 25 m, fast growing, leaves glossy, dark green, 12 cm. Flowers terminal panicles, white, petals joined to form a cap. Fruit ovoid 2–3 cm, purple. Pulp sweet, watery, astringent. Pre-soaking in salt water recommended before eating. Seeds 2. India to Malaysia. Widely cultivated 2, 3, 5, 6

28. **Syzygium grande** (Wight) Walp. SEA ROSEAPPLE

Eugenia grandis Wight.
Tree 20 m. Leaves 20 cm, glossy. Flowers terminal/axil panicles, 3 cm. Fruit globose, pear-shaped, 2–4 cm, green, sweet, dry. 0–1 seed. India, Thailand 2, 3
Syzygium jambolana (Lam) DC., see *Syzygium cumini* (L) Skeels.

29. **Syzygium jambos** (L) Alston. ROSEAPPLE, MALABAR PLUM

Eugenia jambos L., *E. malaccensis* Lour not L., *Jambosa jambosa* Millsp., *Jambosa vulgaris* DC.
Tree 8 m, leaves dark green, 20 cm, flowers terminal, few-flowered, panicles, white. Fruit globose 3–5 cm, cream-yellow, pulp crisp, rose aroma, seeds 0–1, polyembryonic. South East Asia. Widely cultivated 2, 3
Syzygium javanicum, see *Syzygium samagangense.*

30. **Syzygium malaccense** (L) Merrill & L.M. Perry MALAY ROSEAPPLE, POMERAC, JAMBOS, OHIA

Eugenia malaccense L. not Lour., *Jambosa Malaccensis* (L) DC.
Tree 13 m, wood very brittle. Leaves dark green, glossy, to 30 cm. Flowers in cymes on older branches, stamens long, red. Fruit ovoid-pear-shaped, red or yellow, 5–7 cm, pulp crisp, watery, acid-sweet, 1–2 seeds. Polyembryonic in Florida. Not frost hardy. Many forms so it is important to obtain those with best flavour etc. Young plants liable to termite damage. Malaysia. Widely cultivated. 2, 3, 5, 6

31. **Syzygium pycnanthum** Merrill & L.M. Perry. WILD ROSEAPPLE, KOPO

Syzygium densiflorum hort. Brongh & Gris.
Small tree. Used as a rootstock for *Syzygium aquea* and *S. samarangense.* Resistant to termites. East Indies 2, 5

32. **Syzygium samarangense** (Blume) Merrill & L.M. Perry WAXY ROSEAPPLE, JAVA ROSEAPPLE, SEMARANG ROSEAPPLE, WAX JAMBU

Syzygium javanicum Lam not Miq; *Eugenia javanica* Lam, *Jambosa alba* Blume, *Jambosa samarangense* (Blume) DC.
Tree 10 m, leaves dark green, glossy, 20 cm. Flowers paniculate, white, 3 cm. Fruit pear-shaped 3–4 cm, white/red. Pulp watery, sometimes seedless. *S. pycnanthum* as rootstock because of low seed production. Deep soil necessary for good fruits and heavy crops. Malaysia 2, 3, 5

GUAVA

Psidium guajava L.

by D.J. BATTEN

Botany

The guava is a low tree or arborescent shrub, 3 to 10 m high, with a naturally much branched habit. Young shoots are 4-angled, narrowly winged, green and covered with fine hairs. Older shoots are round, wingless, light reddish brown, dull, hairless, with scattered lenticels and bark peeling in thin flakes. Leaves are opposite, 50 to 150 mm long, 30 to 70 mm wide, elliptical, short-petioled, leathery, densely covered with white pubescence when young, later hairless and dark green above, glaucous beneath, or glaucous both surfaces, astringent and bitter with an acrid fragrance when bruised: veins 10 to 25 pairs, sunken above, prominent below; petiole furrowed above, 3 to 15 mm long, with a few glands at the base. Flowers are axillary, borne on new growth from mature branches, pedicelled, solitary or in cymes of two or three, fragrant, hermaphroditic, about 30 mm in diameter, with four or five white petals, numerous stamens and a single thread-like style. Fruits are pendulous and highly variable in size, shape and flesh colour. The calyx lobes tend to persist. Seeds are normally numerous, small, bony, kidney-shaped, flattened, light yellow or yellowish brown, 3 to 5 mm long by 2 to 3 mm wide.

Climatic requirements

The guava originates from the American tropics but can now be found naturalized in many parts of the world in both tropical and subtropical climates. In both tropics it is found at all altitudes up to about 1600 m. Mature plants have been injured and young plants killed at temperatures of –2°C. In hot climates mature trees can recover quickly from occasional frost injury. There appears to be considerable variation between genotypes in sensitivity to cold injury.

No definitive data are available on temperature adaptation. According to Chandler (1958), when the mean summer temperature is less than 16°C, even if the winters are frost free, the guava will not thrive.

The length of the summer growing season is also important. If insufficient warm weather is available for flowering, fruit set and fruit maturation (3.5 to 6 months, depending on genotype and climate), guava cannot be commercially cultivated. Fruit ripening during cold weather often does not develop the proper flavour (Maggs, personal communication).

In a tropical climate it is possible to get two crops per year from guavas, although in India the wet season crop is discouraged in favour of the better quality dry season crop. With natural flowering some guava production is obtained in every month of the year on the Ivory Coast, with peaks of production in March to April and August to October (Fouqúe 1979).

If guavas are grown without irrigation, then the rainfall pattern determines the time of flowering by determining the time of the vegetative flushes which bear the flowers. If rainfall is not consistent following flowering, heavy fruit shedding can occur, and the retained fruit may not size properly (Shigeura 1973).

Soil requirements

Guavas will grow on a wide range of soils from open sand to rather heavy clay (Chandler 1958; Ochse *et al.* 1961). They will survive flooding and continue to grow in soils that are temporarily waterlogged. In the summer of 1978 a spring appeared above the block of guava selections at the Tropical Fruit Research Station, Alstonville, New South Wales, causing the waterlogging of about 20% of the block for

Some guava cultivars

Prepared by D.J. Batten

Variety	Origin	Size†	Shape	Flesh colour	Seeds	Acidity*	Use
Ruby x Supreme	Florida	large	ovoid	white	few	low	dessert
Allahabad Safeda	India	medium	round	white	few	low	dessert
Lucknow–49	India	large	round	white	few, soft	low	dessert
Beaumont	Hawaii	large	ovoid	pink	medium	high	processing
Ka Hua Kula (097)	Hawaii	large	ovoid	pink	few	high	processing
Hong Kong Pink	Hawaii	medium	—	pink	few	low	dessert
Fan Retief	South Africa	medium	—	light pink	medium	medium	dual purpose
11–56	Australia	medium	conical	pink	medium	high	processing
11–56 (T3)	Australia	medium	round	strong red-pink	low-medium	medium	dual purpose

*Low $<$ 0.5% anhydrous citric acid equivalent
Medium 0.5–1.0% anhydrous citric acid equivalent
High $>$ 1.0% anhydrous citric acid equivalent
†Size varies with age, crop load, nutrition, irrigation and other factors: Small $<$ 100 g; medium 100-200 g; large $>$ 200 g.

about 6 weeks and the affected trees did not display any signs of stress. The guava seems to tolerate a wide range of soil pH — from about pH 5 to pH 8. Guavas have some tolerance for salinity (Walker *et al.* 1979).

Genetic diversity

Guava genotypes show great diversity in fruit size, fruit shape, proportion of seed, size and hardness of seeds, flavour, texture of flesh, colour of flesh, aroma, ascorbic acid content, acidity, susceptibility to diseases and pests, susceptibility to fruit cracking, yield, frost tolerance, growth habit (upright/spreading), vigour, time between flowering and harvest, response to defoliation-induced flowering and post-harvest qualities. Some wild seedlings produce very small (40 g), seedy, gritty, musky fruit whereas selected types can produce large (250 g), almost seedless, smooth textured, pleasantly flavoured fruits. The characteristics of processing cultivars differ markedly from those of dessert types (Ito and Nakasone 1973) with processing types often being too acid to be acceptible for consumption as fresh fruit. Dual purpose types appear to show a compromise between processing and dessert requirements.

SOME BETTER VARIETIES.

INDIA

Allahabad Safeeda: fruits round, smooth skinned with white flesh and few seeds, low acidity (less than 0.4%).

Lucknow 49 (Sardar guava): dwarf plant, spreading, profuse bearing. Fruits large (220 g), round with white flesh and few seeds which are soft; acidity low (less than 0.4%).

HAWAII (Ito and Nakasone 1973; Hamilton and Seagrave-Smith 1959).

Beaumont: ovoid shape, pink fleshed, mild flavour, mildly acidic (1.3% total acid), large fruit (220 g), medium amount of seed (1.7% by fresh weight). Processing type. Spreading tree.

Ka Hua Kula (097): ovoid shape, pink fleshed, larger than Beaumont (207 to 277 g), mild guava flavour, mild acidity (1.0 to 1.1%) less seed than Beaumont. Processing type. Vigorous, spreading growth.

Hong Kong Pink: flesh pink and thick with a smooth texture, few seeds, very pleasant flavour, size about 220 g, low acidity (0.2%). Tree a spreading type. Dessert guava.

FLORIDA (Ito and Nakasone 1973)

6363 (Ruby x Supreme): flesh white, thick and sweet, acidity low (0.3%), large fruit (250 g), few seeds. Tree spreading.

SOUTH AFRICA

Fan Retief: sweet, characteristic guava flavour, light pink flesh colour, 2.3% seeds, medium size, considered a dual purpose type in South Africa.

Production season and time of cropping

In warm climates guavas will fruit throughout the year (Fouqué 1979; Shigeura 1973). In the dry tropics the timing of flowering is determined by the availability of water, with intense flowering following the onset of the wet season. Under irrigation, the time of cropping can be manipulated by deblossoming or removal of fruit from trees when fruit is not wanted and by irrigation, fertilizing, and pruning of trees when flowering is desired. Because flowers are borne on new growth, anything which stimulates growth from mature wood will stimulate flowering — thus defoliation with urea sprays has been shown to be effective in stimulating guava flowering (Shigeura *et al.* 1975; Chapman *et al.* 1979).

Under conditions where the dry season is ill defined, the harvesting of guavas is protracted, the control of pests and diseases is difficult, and harvesting costs are high. Under such conditions manipulation of flowering by defoliation can markedly reduce costs of production by contracting the harvest period. Furthermore, deliberate staggering of defoliation times for different blocks of guavas should prevent seasonal gluts in supply.

Under subtropical conditions such as prevail in Florida, and latitudes 25 to 30°S on the east coast of Australia, guavas usually flower naturally in spring-early summer and the fruit matures in late summer-autumn. However, wild guavas sometimes flower in autumn with the fruit maturing in spring.

When the winter is severe enough to defoliate trees, flowering begins as soon as it is warm enough for new growth to occur and it is difficult to envisage means of exercising control over flowering in such climates. In India trees defoliate during the dry season in some areas and this acts in a similar manner to stimulate intense flowering following the onset of the wet season.

Guavas begin fruiting at an early age. Two-year-old clonal trees defoliated with 25% urea solution have yielded an average of 61 kg of fruit per tree at Alstonville. Three-year-old trees defoliated with ethephon-gibberellic acid-urea have produced an average of 97 kg of fruit per tree.

Yield potential

With yields of 60 kg or more fruit per tree possible from 2-year-old trees, the yield per hectare at a commercial plant spacing of 6 m x 2.5 m (667 trees per hectare) could be 40 tonnes or more. Such yields are already possible. The limits have hardly been tested yet. The main problem with hanging such heavy crops on young trees is the limb breakages. It may not be possible to get such high yields from trees being trained for mechanical harvesting because of the amount of pruning needed on these trees.

Market acceptability

Dessert fruit

Good quality dessert guavas find a ready market in Asian countries. We do not know the domestic market potential. It is important that only the best quality fruit be marketed. Market development could be given a severe set-back if fruit are marketed from processing selections, wild guavas or poor quality dessert types. Controlled cropping should allow the marketing of guavas when there are lulls in the availability of other fresh fruits.

Processing fruit

The current Australian juice market is approximately 300 ML of drink per annum. At yields of 50 tonnes per hectare, 100 hectares of guavas would make enough 20% drink to supply 6% of the total juice market. It would seem therefore that development based solely on a domestic market would be limited. Guavas can also be marketed as chunks in syrup and as guava cheese and other confectionery products.

Export potential

At this stage, the prospects for exports of processing guava look good. How long this will last no one knows. Several things are in our favour:

- Asian countries have juice industries based on dessert varieties. The juice is grossly inferior to that which can be made from the specially developed processing types we have here.
- Production technology is advanced here compared with cheap labour countries.
- UHT, pasteurized and depectinized concentrate, could be made here with little problem, thus markedly reducing the cost of transport and making the Australian product more acceptable as a 'healthy' food (no preservatives).

There does not appear to be much prospect for exports of fresh guava fruit because:

- the value of the fruit is fairly low;
- the fruit does not ship well and has a limited post-harvest life; and
- exported fruit would have to be fumigated to be acceptable to most foreign markets (because of fruit fly risks).

Cultural details

Propagation

Guavas are easy to propagate. The advantages of clonal propagation may not outweigh the added cost over seedlings for processing guavas because most seedlings retain the general characteristics of the parent tree. See Nakasone (1973) for techniques.

Seedlings. The medium must be pasteurized to avoid damping off problems. Once 50 mm high, seedlings are quite resistant to damping off and can be transplanted into pots without special hygiene precautions. Seedlings are prone to root binding and normal procedures should be followed to avoid this problem. If the temperature is high (30°C or more), seed germinates in about 15 days, usually with more than 90% of fresh seed germinating. Seed appears to store satisfactorily for up to a year or more. Selection 11–56 has about 100 seeds per gram weight.

Budding. The best method of propagation. Guavas can be budded by the patch-bud or the modified (Forkert) patch-bud technique (Hawaii). We have had excellent success with chip budding at Alstonville in February. Buds need to be wrapped completely with PVC budding tape. A small hole cut in the tape allows the new shoot to emerge. When the shoot is 60 mm or more long the tape is removed and the top of the stock cut off just above the new bud.

Scionwood for budding or grafting is prepared 10 to 14 days before it is needed by removing the leaves from the branch. This causes the axillary buds to enlarge. Wood that is shedding its rough bark or has shed its bark and is without leaves gives excellent results.

Grafting. The side-wedge is occasionally used in Hawaii.

Marcottage. The usual technique can be used for small numbers of plants.

Cuttings. Softwood cuttings will root if treated with 2000 to 4000 ppm IBA and placed under properly regulated intermittent mist, in a free-draining medium, with 28 to 30°C bottom heat. The cuttings should consist of two pairs of leaves and the terminal cluster. The cuttings should be green in colour not turning brown (Maggs, personal communication). In Hawaii, shaded polythene frames have substituted satisfactorily for mist. The biggest problem with rooted cuttings is their susceptibility to wind damage in their first 2 or 3 years (Nakasone 1973).

Management

Planting. North-south rows, 2.5 to 4.0 m within-row-spacing and 6 to 7.7 m between-row-spacing. The higher density of 667 trees per ha (2.5 x 6.0 m) will maximize early returns, although in the long term some trees may need to be removed. Container-grown trees should be checked for root binding and the bent root growth cut off.

Wind-breaks. Not as important as for macadamia in Hawaii (Shigeura 1973) but still very desirable. Loaded trees break easily in windy conditions. Cold winds will restrict growth, flowering, and fruit development.

Training. The proposed method of harvesting determines the type of tree structure desired. For tree shaker/catcher harvesting, the trunk must be clean up to about 80 cm from the ground and the laden branches must be clear of the ground. For hand harvesting it is not important to get the canopy off the ground. Since guavas tend to grow in a shrubby manner, continual attention is needed in young trees to maintain a clean trunk.

In either case, the aim should be to encourage the development of woody (flower-bearing) twigs close to the main framework. This is done by severely pruning long shoots protruding from the dome of the young tree. Small branches within the crown of the tree should never be removed — these will bear fruit.

Crop regulation. Flowering can be induced by pruning and by defoliation. Defoliation can be achieved by injuring leaves with a 25% urea spray and this causes intense flowering (Chapman *et al.* 1979). Ethylene-releasing chemicals can also be used — 1600 ppm ethephon with 5% urea and wetting agent has worked well at Alstonville and is much easier to prepare than the 25% urea solution. Desiccant herbicides might also be an economical way of defoliating trees. Any pruning should follow defoliation and trees should be fertilized and irrigated to encourage a strong flush of growth.

Guava growers in a number of Asian countries encourage flowering by tying branches to a horizontal position. Deblossoming is used in India to discourage the wet season crop in favour of the dry season crop.

Irrigation

Irrigation is essential to properly regulate flowering and to achieve high yields. Water stress during fruit development will cause fruit shedding and dramatically reduce yields.

Fertilization

No leaf analysis standards are available. Little research has been done on the nutritional requirements of guavas. Guavas will survive in low fertility soil but they respond well to heavy applications of fertilizer. Established trees seem to be fairly resistant to fertilizer burn.

In South Africa the recommendation is 250 g N, 100 g P, and 500 g K per mature tree in spring before flowering (induced by pruning) and two applications of 60 g N each, one at commencement of growth after pruning, and the other when the fruit begins to swell. Cracking of fruit is said to be associated with high nitrogen levels (Cull 1977).

In Hawaii, the 'rule-of-thumb' of 200 g of mixed fertilizer (N:P:K ratio of 9:9:9) per cm of trunk diameter per year is used.

Timing of fertilizer application is important as part of the cropping control strategy. Fertilizing prior to flowering promotes growth and thus flowering (Hamilton and Seagrave-Smith 1959). However, excessive fertilizer at the time of flowering might prolong the flowering period. More experimental data are required for the effect of fertilizing on flowering, fruit set, fruit growth, fruit splitting, and the length of the harvest period.

Pests

- Fruit fly. Relatively easy to control in controlled cropping situations with cover sprays or bait sprays.
- Scale insects. In South Africa and Hawaii, insects are successfully controlled by natural predators, ants. Defoliation reduces the level of infestation.
- Thrips. One selection at Alstonville has been severely affected by thrips which cause scarring of the skin of the fruit and subsequent cracking.
- Leaf-eating pests. Various lepidopterous larvae can severely damage young plants. Some level of activity might be beneficial in performing a tip-pruning service. *Monolepta australis* has caused serious defoliation in one planting near Alstonville. Other species of leaf-eating beetles cause serious damage in northern Queensland.

- Birds and fruit bats can damage fruit. However the highly acid types will possibly not be as attractive to fruit bats as wild guavas or the sweet dessert types.

Diseases (see William 1975)

Very little research has been done on guava diseases in Australia.

- Blossom end rot. Dark brown circular spots at the blossom end of immature fruit. Serious in rainy weather with up to 25% or more of fruit affected in India where it is controlled with 0.2% Dithane Z-78 sprays at monthly intervals. There may be physiological as well as pathological causes of blossom end rot.
- Fruit canker *(Pestalotia psidii)*. Circular, dark coloured, raised corky growths on fruit. Affected immature fruit does not ripen properly. In India Dithane Z-28 (0.2%) or Difolatan (0.3%) are used at monthly intervals for control.
- Anthracnose *(Colletotrichum gloeosporioides)* causes circular scabby spots on fruit which may be pushed out in a complete plug. Also causes necrotic ash-grey lesions on the tips and margins of leaves. Pruning of diseased twigs and destruction of fallen fruit helps. Control measures for fruit canker also control anthracnose in India.
- Mucor rot *(Mucor hiemalis)* produces water-soaked areas on fruit. Within a week the whole fruit is affected and is covered by a yellowish fuzzy mass of fungal fruiting bodies and mycelia. Wounding is necessary for infection. Sanitation drastically reduces the disease incidence (Ito *et al.* 1979).
- Guava wilt. Thought to be caused by *Fusarium solani* in India and *Myxosporium psidii* in Taiwan. Varieties vary greatly in susceptibility — most red-fleshed cultivars are resistant and most white cultivars are susceptible (Singh *et al.* 1977). Wilt is a serious disease in northern India.
- Leaf spots. *Cephaleuros virescens, Cercospora psidii,* and *Puccinia psidii* cause leaf lesions in Hawaii (Martinez 1973).
- Wood rots. Various pathogens, including *Clitocybe tabescens,* cause wood rots in guava (Martinez 1973).
- Nematodes. Guava is susceptible to *Meloidogyne incognita* and *M. arenaria. Psidium friedrichsthalianum,* which is resistant to *Meloidogyne,* could be tried as a rootstock.

Harvesting methods

If fruit harvesting intervals exceed 4 days in warm weather loss of overripe fruit becomes a problem. The usual method is hand harvesting. Field bins should not be more than 25 cm deep or fruit will be crushed.

Mechanical harvesting is being investigated in Hawaii with a machine based on the concept of the Cornell Apple Foliage Shaker (Cull 1977).

Catching nets have been used experimentally in Puerto Rico. Fruit falls from the tree and is collected every 7 days without problems (Lopez Garcia and Perez Perez 1977).

It will probably be possible to shorten the harvest period by stimulating ripening with growth regulators.

Transport characteristics

The fresh fruit is highly perishable and must be marketed promptly. Ripe fruit can be stored for up to 2 weeks at 4°C (Hamilton and Seagrave-Smith 1959). Waxing has extended shelf life from 3 to 6 days in trials in India (Garg *et al.* 1976). Benomyl is effective against two of the major post-harvest pathogens in India (Singh and Bhargava 1977).

Cultivars vary in their susceptibility to bruising and rotting. Allahabad Safeeda is much more susceptible to rots than apple guava (a red-skinned cultivar).

Processing characteristics

Guavas may be used to make a delicious fruit juice drink, jams, jellies, guava 'cheese', or pieces can be canned for consumption as dessert fruit.

Fruit for juice production needs to have a high total soluble solids (~10° Brix) and a good flavour. A strong pink colour is preferred. Stone cells can be removed from processing puree with a fine screen finisher but dessert fruit for canning must have a smooth texture (free of stone cells).

References

Chandler, W.H. (1958), 'Evergreen orchards'. 2nd edn, Lea and Febiger, Philadelphia, p. 326.

Chapman, K.R., Saranah, J., and Paxton, B. (1979), Induction of early cropping of guava seedlings in a closely planted orchard using urea as a defoliant. *Australian Journal of Experimental Agriculture and Animal Husbandry* **19,** 382.

Cull, B.W. (1977), Report on overseas study tour, sub-tropical fruit. Queensland Department of Primary Industries, Horticulture Branch.

Fouqué, A. (1979), Quelques observations sur les goyaviers. *Fruits* **34,** 767.

Garg, R.C., Ram, H.B., and Singh, S.K. (1976), A note on the effect of post-harvest treatment with indol-butyric acid and wax emulsion on the storage behaviour of guavas *(Psidium guajava* L.*). Progressive Horticulture* **8,** 85.

Hamilton, R.A., and Seagrave-Smith, H. (1959), Growing guava for processing. University of Hawaii Extension Bulletin 63.

Ito, P.J., and Nakasone, H.Y. (1973), Important guava cultivars for Hawaii, *in* Hawaii Agricultural Experiment Station and Co-operative Extension Service Miscellaneous Publication 111, p. 19.

Ito, P.J., Kunimoto, R., Ko, W.H. (1979), Transmission of mucor rot of guava fruits by three species of fruit flies. *Tropical Agriculture* **56,** 49.

Lopez Garcia, J. and Perez Perez, R. (1977), Effect of pruning and harvesting methods on guava yields. *Journal of Agriculture of the University of Puerto Rico* **61,** 148.

Martinez, A.P. (1973), Guava diseases, *in* Hawaii Agricultural Experiment Station and Co-operative Extension Service Miscellaneous Publication 111, p. 8.

Nakasone, H.Y. (1973), Guava propagation in Hawaii, *in* Hawaii Agricultural Experiment Station and Co-operative Extension Service Miscellenous Publication 111, p. 16.

Ochse, J.J., Soule, J.J. Jr, Dijkman, M.J., and Wehlburg, C. (1961), 'Tropical and sub-tropical agriculture'. Macmillan, N.Y., p. 686.

Shigeura, G.T. (1973), Culture and management of guava, *in* Hawaii Agricultural Experiment Station and Co-operative Extension Service Miscellaneous Publication 111, p. 10

Shigeura, G.T., Bullock, R.M., and Silva, J.A. (1975), Defoliation and fruit set in guava. *HortScience* **10,** 590.

Singh, A.P., and Bhargava, S.N. (1977), Benlate as an effective post-harvest fungicide for guava fruit. *Indian Journal of Horticulture* **34,** 309.

Walker, R.R., Kriedemann, P.E., and Maggs, D.H. (1979), Growth, leaf physiology and fruit development in salt stressed guavas. *Australian Journal of Agricultural Research* **30,** 477.

William, D. (1975), Diseases of guava, *in* 'Advances in mycology and plant pathology'. S.P. Raychaudhuri, *et al.* (eds), New Delhi. Professor R.N. Tandon's Birthday Celebration Committee. p. 179

Further reading

Bourke, D. O'D. (1976), Propagation of tropical fruit trees, *Psidium guajava,* guava, *in* 'The propagation of tropical fruit trees'. R.J. Garner, *et al.* (eds), Horticultural Review No. 4, Commonwealth Bureau of Horticulture and Plantation Crops, East Malling, U.K., p. 530.

Nakasone, H.Y., Brekke, J.E., and Cavaletto, C.G. (1976), Fruit and yield evaluation of ten clones of guava *(Psidium guajava).* Hawaii Agricultural Experiment Station Research Report 218.

Singh, U.R., Dhar, L., and Singh, G. (1977), Note on the performance of guava cultivars and *Psidium* species against wilt disease under natural field conditions. *Haryana Journal of Horticultural Sciences* **6,** 149.

Teaotia, S.S., and Phogat, K.P.S. (1971), Effect of rootstocks on growth, yield and quality in guava *(Psidium guajava). Progressive Horticulture* **2,** 37.

FEIJOA (PINEAPPLE GUAVA) by D.J. BATTEN
Feijoa sellowiana O. Berg.

Valuable as an ornamental, the feijoa has also been used as a wind-break species in New Zealand for many years.

Botany

Feijoa sellowiana O. Berg. belongs to the family Myrtaceae. It is a shrubby tree, about 5 m high at 10 years of age. Leaves are small, about 50 mm long, glossy bluish green on the upper surface, silver-grey on the lower. Flowers are attractive, about 30 mm in diameter. The four petals are fleshy, edible, white outside, purplish within. Style and numerous, conspicuous stamens are red. Fruit are round, oval or oblong, 20 to 80 mm long, dull green with a whitish bloom. Skin is thin, inedible, enclosing a whitish layer of granular flesh and a central mass of translucent, jelly-like pulp containing 20 to 30 minute seeds.

Climatic requirements

Feijoa is very hardy — quite drought resistant and frost tolerant (–9°C). It grows well at Santa Barbara, California, without irrigation but needs as much water as citrus for best results. Feijoa has not succeeded in moist tropical regions (Popenoe 1920). According to Chandler (1958), the best flavour is developed in the cool summers of coastal California rather than the hot summers of interior southern California. The plant appears to be adapted to a mild climate, free from extremes of heat and cold. Although the plant can tolerate –9°C, –1°C for more than an hour can damage the fruit.

Soil requirements

The feijoa tolerates many types of soils but does best on well-drained, fertile soils. It might not tolerate alkaline soils (Popenoe 1920).

Genetic diversity

The present industry in New Zealand is based on seedlings and these exhibit a wide range of fruit and bush types. The main selection criteria for varieties are: large fruit size (120 to 150 g) and blocky shape; consistent fruit size and shape; good flavour; large volume of seed-pulp; self-fruitfulness; yield; bush habit (upright rather than sprawling); resistance to bruising; post-harvest life; resistance to browning when cut; and short harvest period.

The best varieties at this stage are (Endt 1978; Jones 1975):

- Mammoth: fruit large, oval, smooth skinned, reasonable yield. Ripens mid-April in New Zealand.
- Triumph: fruit large, oval, skin smooth. Yields heavily. The best selection so far.
- Other varieties: Coolidge has variable fruit size and shape but is a heavy cropper. Choiceana, Superba, Large Oval, and Chapman are all inferior in quality.

Even Mammoth and Triumph show some self-incompatability which is seen in a variable degree of pulp development. Cross-pollination by bees ensures full pulp development. Thus both Mammoth and Triumph should be interplanted to ensure cross-pollination. Selection for self-fertile types with consistent pulp development is needed (K. Patterson, DSIR, New Zealand personal communication).

Production season and time of cropping

In New Zealand, feijoas begin flowering in December and the fruit ripens in April-June. In Florida, flowers appear in late April, in southern California in May, and in the cool San Francisco Bay district in June. Fruit ripens 4.5 to 6 months after flowering in southern California and 5.5 to 7 months after flowering in the San Francisco Bay district (Chandler 1958).

There have apparently been no efforts made to induce flowering to control the cropping cycle.

Yield potential

A 2-year-old vegetatively propagated Triumph or Mammoth should yield about 15 kg per tree (R. Bell, Victorian Department of Agriculture personal communication). The yield will depend on how quickly the plants grow which is determined by both climate and management.

Feijoas grow more slowly than guavas and the yields will be much less.

Market acceptability

The flavour of feijoas appeals to many people. The flavour resembles a blend of strawberry and pineapple. Feijoas keep well if unbruised. The only problem will be in educating consumers about the stage of ripening at which the fruit should be eaten. One problem concerns the lack of external signs to indicate the degree of seed pulp development. Fruit can look similar but differ markedly in degree of pulp development.

Export potential

Fruit is being exported from New Zealand. Exports in 1975 returned about $2.70 per tray net to growers. Ultimate market size is unknown.

Cultural details

Propagation

Seedlings are easy to raise. Dry seed keeps for a year or more. Seed germinates rapidly without stratification. Seedlings do not come true to the parent and can be completely unfruitful.

Cuttings are probably the best method of propagation. Three-node semi-hardwood cuttings taken in early winter are most successful. Mist, bottom heat (21°C), and IBA-treatment (200 ppm or Seradix No. 3) are necessary.

Grafting. One-year-old seedlings can be whip-grafted if kept at a high humidity.

Management

Spacing should be about 2 m to 3 m within rows and 5 m between rows, this spacing results in a hedgerow development. Wind protection is essential. Wind causes rubbing and marking of fruit.

Pruning/training. Growth close to the ground is removed to obtain some clear trunk. Little detail pruning is practised. Mechanical hedging may be possible as the flowers are borne on current season's growth.

Fruit thinning. Some growers remove small or mis-shapen fruit in attempts to increase fruit size.

Irrigation

Feijoas respond well to summer irrigation (during flowering and fruit development). Precise requirements are not known. Irrigated trees produce larger fruit than unirrigated trees.

Fertilization

Feijoas get by with little fertilizer but the optimm has not been determined.

Pests and diseases

In New Zealand the following pests affect feijoas: Greedy scale infests young branches and fruit; Hard wax scale *(Ceroplastes sinensis)* infests young branches and leaves, and persists on woody stems; Bag moth larvae tunnel into stems; Light brown apple moth larvae eat young leaves — the fruit skin can be damaged; Black scale *(Saissetia oleae)* has been noted as a principal pest in California; Fruit fly will be a major pest of feijoas in much of Australia.

Few diseases have been identified. *Sphaceloma* fungus infests leaves and sometimes fruit in New Zealand.

Harvesting methods (Endt 1978)

Fruit falls to the ground when it is mature. Such fruit bruises and rots and is unsuitable for the fresh market. Fallen fruit can be sold for processing in New Zealand.

Fruit for the fresh market must be picked from the tree. A subtle colour change indicates maturity and the lack of a distinct maturity indicator can be a problem. However, mature fruit when handled gently will break from the tree.

Fruit should be picked daily. It may be possible to develop a system of catching nets to aid harvesting.

Marketing, storage, transport characteristics

The fruit is tender and requires careful and prompt handling. It should be placed in cool storage as soon as possible after picking. Export fruit in New Zealand is packed in polystyrene trays and despatched by air in units of three. Moulded plix trays of 30, 35, 40 and 45 are used. Consumers need to be educated about how to use feijoas.

Fruit picked from the tree can be held in cool storage at 4 to 5°C for up to 3 months. The fruit will then ripen at room temperature and keep for 7 to 14 days. (Ethylene triggers ripening. Fruit falling to the ground is thought to ripen quickly because of the wound [bruising] — induced ethylene production). It might be possible to ship fruit by surface transport by packing ethylene absorbents with the fruit.

Processing characteristics

The expected yields and production of processing fruit would not be very profitable. However, a proportion of the crop will not be fit for marketing as dessert fruit (fallen, mis-shapen, small) and the fruit is suitable for processing.

The skin can be easily removed with a hot alkali solution.

The strong flavour of the skin disappears when heated so fruit can be canned whole.

The fruit juice is too strongly flavoured alone but blends well with other juices.

Feijoa juice apparently makes excellent wine.

Frozen feijoa pulp stores well.

References

Chandler, W.H. (1958), 'Evergreen orchards'. Lea and Febiger, Philadelphia.

Endt, D.J.W. (1978), The feijoa — a fruit of promise. *Orchardist of New Zealand* **51**, 341.

Jones, D.L. (1975), Report on overseas trip — the kiwifruit (Chinese gooseberry) industry in New Zealand and notes on feijoa production. Scoresby Hort. Res. Stn., Fern Tree Gully, Victoria.
Popenoe, W. (1920), 'Manual of tropical and sub-tropical fruits'. Macmillan, N.Y.

Further reading

Macpherson, N. (1952), Feijoas have a variety of culinary uses. *New Zealand Journal of Agriculture* **84,** 336.

Chapter 19

OXALIDACEAE

CARAMBOLA or **FIVE CORNER** by M. SEDGLEY
Averrhoa carambola L.

BILIMBI or **PICKLE FRUIT**
Averrhoa bilimbi L.

Most of the following description applies to *A. carambola* as more information is available than for *A. bilimbi.*

Botany

A. carambola: The evergreen tree native to South East Asia grows to 5 to 12 m high and has compound leaves with between 2 and 11 leaflets each 2 to 9 cm long. The flowers are pink, perfect, 8 mm long and occur in axillary and terminal clusters. The fruit is a fleshy berry, 8 to 12 x 3 to 6 cm, acutely 5 angled and star shaped in cross-section. It is yellow when ripe and has 10 to 12 ovoid seeds, 1 cm long.

A. bilimbi: The evergreen tree native to South East Asia grows to 15 m high and has compound leaves with between 20 and 40 leaflets each 5 to 10 cm long. The flowers are red, 2 cm long and occur in clusters from the trunk and older branches. The cylindrical fruits are 5 to 7.5 cm long and faintly 5 angled. They have small seeds 8 mm long and embedded in pulp.

Climatic requirements

Both species require tropical or warm subtropical lowland conditions. The trees cannot establish in areas prone to frost as the young trees are damaged. Older trees are more frost tolerant than young trees, but can withstand low temperatures for only short periods.

Soil requirements

A. carambola thrives on almost any type of soil provided it is reasonably well drained. It does best on a moderately acid soil of pH 5.5 to 6.5.

Genetic diversity

There is variability in the fruit of *A. carambola* from sour to sweet. Many selections and established cultivars with acceptable yield and disease tolerances are available. Named cultivars include Demak Blimbing, Wheeler, Kwangtung, Starking, Mih Tao, Maha, Dah Pon, Pei Sy Tao, Newcomb, Thayer Lengbak, Jurong and Taen Ma. The best cultivars are Wheeler and Kwangtung.

There is considerable scope for the development of improved varieties of *A. carambola.* Knight (1964) has shown that controlled hand pollination is both feasible and practical and seedlings often bear in their second year allowing rapid selection for fruit quality.

With *A. bilimbi,* sweet and acid forms exist but no cultivars have been developed.

Some carambola cultivars

Cultivar name/number	Origin	Vegetative characteristics	Fruit characteristics	Pest susceptibility	Comments
B1 Yong Toh Yin	Malaysia	—	Large, yellow, sweet, good flavour	—	Good keeping quality
B2 Maha 66	Malaysia	slow growing tree	Medium-large. Pale greenish yellow, sweet, good flavour, can be eaten immature	Susceptible to fruit fly even when immature	Heavy bearer Best Malaysian cultivar
B10 Ching Sing Keow	Malaysia	vigorous	Large ovate fruit which narrows toward the base. Yellow orange, sweet, good flavour. Many seeds.	Moderately resistant to fruit fly	Heavy bearer Superior cultivar
B11 Chan Yong 1	Malaysia	—	Large, yellow-green, sweet.	—	—

Information on Malaysian cultivar translated by Dr. R. Winoto Suatmadji from a paper by Mohd. Nor bin Ibrahim, Mohamed bin A. Shukor and Lye Tuck Thye.

Production season and time of cropping

Both species flower and fruit throughout the year. In Australia the main flowering months are May to June and the main cropping season is late winter.

Yield potential

In Florida *A. carambola* trees start to bear at 3 years and a mature tree will produce 50 to 150 kg of fruit per tree per year. Mature trees in southern Australia have produced over 800 kg in a 9 month period. The trees are long lived and consistent bearers.

Market acceptability

Fresh *A. carambola* fruit has a potentially high market acceptability (Wagner *et al.* 1975). The ripe fruit is attractive and in the sweeter cultivars has a pleasant flavour. It is high in vitamin C containing between 14 and 90 mg per 100 g of fruit depending on the cultivar. Carbohydrate levels are 8 to 15% and there is some vitamin A. A disadvantage is the oxalic acid content of up to 0.51%. This can probably be reduced by breeding and selection. The fresh fruit has high acceptability in its limited markets in the U.S.A. and Sweden. The fruit can be eaten when green.

A. bilimbi fruit is very acid, even in the sweeter forms and is not suitable for the fresh fruit market.

Export potential

A. carambola can be exported as fresh fruit as small batches have been exported from Florida to Sweden. Australian produced fruit would probably be acceptable to the Asian markets in the fresh, and also possibly the preserved, form. The fruit is more familiar in Asia than in Australia and so would encounter less consumer resistance.

Cultural details

Propagation

A. carambola can be successfully propagated by seed and grafting. Budding, cuttings and air layering are possible but less reliable. Top working is also successful. Less work has been done on *A. bilimbi* which can be propagated by seed and air layering (Tidbury 1976).

Management

A. carambola trees can be planted at any time of the year although September to April is recommended for Australia. Trees should be planted at approximately 6 m spacings. The tree is slow growing and long lived and needs only occasional pruning when mature. It has poor drought and salinity tolerance but is severely damaged by flooding. It is resistant to wind damage and thus is a potentially very useful tree. However, wind does cause heavy fruit fall. *A. carambola* is a heterostylous species with cross- and self-incompatibility (Knight 1965). Interplanting of cultivars and the provision of insect pollinators would be required. However, in the cultivar Golden Star the incompatibility mechanism has broken down. The tree is thus self-fertile although insect pollinators are still required.

A. bilimbi is less wind tolerant than *A. carambola* and is reportedly self-fertile.

Irrigation

Irrigation is recommended particularly following planting and during dry periods. Adequate water is required.

Fertilization

The Florida recommendations (Malo and Campbell 1972) are for 0.25 kilos of fertilizer around each tree every 2 months up to 5 years and thereafter 1 to 3 applications per year of 0.5 kilos per cm of trunk diameter. The fertilizer should contain N:P:K:Mg in a 1:1:1:0.5 ratio. On alkaline soils supplementary Fe, Zn and Mn are required, usually as iron chelate and foliar sprays

Pests and diseases

No pests or diseases of *A. carambola* have been described in Florida apart from occasional bird damage. Fruit fly is reported to be a potential pest in Darwin and a fruit rot caused by *Cladosporium herbarum* has been described from India (Sharma and Khan 1978).

Marketing characteristics

A. carambola could be marketed either as fresh fruit or in the processed form. *A. bilimbi* is very acid as fresh fruit and could be marketed only in the preserved and pickled form.

Storage characteristics

The best temperature for storage of *A. carambola* fruit is 10°C. The fruit is reportedly sensitive to adverse storage temperatures.

Transport characteristics

Opinions appear to differ on the transport characteristics of *A. carambola* fruit. However, the fact that the fresh fruit is successfully transported by sea from Florida to Sweden suggests that any problems are minimal.

Processing characteristics

There is no commercial processing of either species at present as there are no sizeable plantations anywhere in the world. However, both species are successfully processed domestically. *A. carambola* can be dried, juiced or made into jams, jellies and drinks and *A. bilimbi* into pickles, curries and preserves.

References

Knight, R.J. Jr (1964), The carambola in South Florida. Rare Fruits Council of South Florida, Occasional Paper 1, 3.

Knight, R.J. Jr (1965), Heterostyly and pollination in carambola. *Proceedings of the Florida State Horticultural Society* **78**, 375.

Malo, S.E. and Campbell, C.W. (1972), The carambola. University of Florida Fruit Crops Fact Sheet No. 12.

Sharma, N. and Khan, A.M. (1978), Fruit rot in *Averrhoa carambola* L. *Geobios* **5**, 48.

Tidbury, G.E. (1976), *Averrhoa* spp. — carambola and bilimbi, *in* 'The propagation of tropical fruit trees'. R.J. Garner and S.A. Chaudri (eds), Commonwealth Agricultural Bureau.

Wagner, C.J. Jr, Bryan, W.L., Berry, R.E. and Knight, R.J. Jr (1975), *Proceedings of the Florida State Horticultural Society* **88**, 466.

Chapter 20

PALMAE

subfamily Calameae

SALAK by D.H. MAGGS
Salacca edulis Reinw.

Botany

Salacca edulis is an understorey palm native to Indonesia, especially Java, Bali and Modeno, and widely cultivated in the wetter parts of the Indo-Malayan region.

It is normally short and dumpy, but if it grows tall and topples over, it will send out roots where the trunk comes in contact with the ground. It may also produce basal suckers but does not form large clumps. Feather-leaves are up to 7 m long; the margins of midrib, petiole and leaflet are set with long sharp spines; leaflets are up to 70 cm long and dark green.

Male and female flowers are on separate plants. Male flowers are closely packed in finger-like spadices borne 4 to 12 together in bunches. Female flowers are larger than the males, on shorter spadices borne 1 to 3 together. No information is available on mode of pollen transfer.

Fruits are in tight, globose bunches, round, 2.5 to 10 cm x 5 to 8 cm across. They are covered with regularly arranged scales developing from the skin of the fruit (pericarp) giving it the appearance of a small portly pine cone. The fruits enclose a soft, translucent pulp which tastes of pineapple-cum-banana when ripe. Fruits contain 1 to 3 blackish nuts. The kernel is about 1 cm in diameter.

The genus contains about nine other species of which at least three may be mistaken for *S. edulis.* These are:

- *S. affinis* Blume in Malaysia (that is, the east-west zone immediately north of Indonesia).
- *S. conferta* Griff. Malaysia. A swamp species with very sour fruits.
- S. wallachiana Mart. Malaysia and Siam. An extremely dumpy species.

Climatic requirements

The species requires tropical rainforest conditions with an annual rainfall of 2000+ mm, 90 mm in the driest month. The temperature range required is 22 to 32°C, with an elevation of 1 to 300 m. *S. edulis* is said not to flower in the Singapore Botanic Gardens although the climate is within these limits.

Soil requirements

S. edulis requires a free-draining soil with a high organic content.

Genetic diversity

Cultivars have been selected from the wild forms, varying in fruit size and colour. Seedless and monoecious forms occur, especially in Bali. Seeds from 1-seeded fruits produce mostly male plants and from 2-seeded fruits mostly female (Ochse *et al.* 1961). The species is in the Australian nursery trade.

Production season and time of cropping

The species bears fruit throughout the year.

Export potential

Preserved fruits are exported to Singapore.

Cultural details

Propagation

Seed. Four to five seeds sown under an established banana nurse clump, or raised in a shaded nursery and transplanted during the more rainy periods.

Suckers. Especially those regenerating from fallen plants.

Management

Maintenance of shade and organic litter, and weed control should be undertaken.

Irrigation

In Australia irrigation would probably be necessary during dry spells, lasting more than 10 days. Intermittent mist to maintain humidity might also be tried.

Fertilization

A response to N, P would be expected.

Marketing characteristics

Ripe fruits cannot be kept for more than 3 to 4 days. The fruit is mostly sold fresh in local markets. Some fruits preserved in sugared brine are exported to Singapore.

Processing characteristics

Fruits are canned in sugared brine (Ochse *et al.* 1961). Green fruits may be pickled.

Reference

Ochse, J.J., Soule, M.J. Jr, Dijkman, M.J., and Wehlburg, C. (1961). 'Tropical and sub-tropical agriculture'. MacMillan, New York.

Further reading

Burkill, I.H. (1966). 'Dictionary of the economic products of the Malay Peninsula'. Min. Ag. & Coop., Kuala Lumpur.

subfamily Phoenicae

DATE PALM by D.H. MAGGS

Phoenix dactylifera L.

Botany

The genus *Phoenix* (17 species) is spread across the drier tropics from the Canary Islands to south-west China. The date palm probably originated in the Persian Gulf region and is now widespread as an oasis species throughout the driest areas. It has the largest fruit in the genus.

It reaches a height of up to 30 m with annual increments reaching 50 cm. The trunk is covered with the remains of dead leaves which must be removed during cultivation. It suckers at the base when young and sometimes forms dense thickets. Leaves are pinnate, 3 to 4 m long, terminal and basal leaflets forming robust spines. Up to 30 leaves are produced annually each lasting 3 to 7 years; there are 2, sometimes 3, leaves to a whorl (divergence angle 137.6°).

Throughout the life of the palm numerous roots develop from the base of the trunk. These are about 1 cm in diameter and subdivide repeatedly to form a widespread fibrous mat capable of blocking field drains. The roots have a hard, persistent skin and a central conduction strand; the space between is filled with short-lived parenchyma.

The species is dioecious and wind pollinated. The young inflorescences are enclosed in a strong spathe which splits open at anthesis. They consist of a bundle of 25 to 100 spikelets 20 to 80 cm long borne on a stout main axis. This axis elongates rapidly as the fruits enlarge so that the mature bunch hangs down well clear of the foliage. Seven to twelve inflorescences are produced each year.

On the male spikelets the flowers are crowded together but on the females they are separated into clumps of 1 to 3 flowers. Receptivity lasts 10 to 15 days. The female flower has 3 carpels. Normally, only one carpel develops after pollination. Unpollinated flowers frequently develop into a cluster of 3 seedless fruits. The fruit is a drupe. When the fruit is plump-ripe, the edible mesocarp contains up to 48% sugars and 40% water and gradually dries out to 20% water. Green fruits are astringent because of soluble tannins which precipitate as the fruit ripens.

Date varieties may be divided into 3 groups: dry (bread) dates whose mesocarp sugar is 100% sucrose; soft dates with 95+% fructose plus glucose; and semi-dry dates usually with 80 to 90% fructose and glucose, but sometimes less. World trade is mostly in soft and semi-dry varieties, while locally, the bread dates are extensively traded.

Climatic requirements

Date palms grow in many parts of Australia and a commercial grove still operates in Alice Springs. They have fruited on research farms at Miles and Barcaldine, Queensland.

They require: i) high temperature during active growth — mean maximum temperature for January and February of 35°C. ii) low humidity and rainfall while the fruit ripens and softens — for early varieties, a median rainfall less than 25 mm in February, March; for late varieties less than 10 mm in April, May. iii) a nine-month growing season with mean monthly temperature in August, over 12°C, in September, over 15°C.

Useful data sources: Bureau of Meteorology, 1974, 1975. The 90 percentile isohyets in Bureau of Meteorology (1975) give an indication of the reliability of the climate.

Soil requirements

Dates are tolerant of high water-tables, periodic flooding, salinity, alkalinity and drought, but best results are obtained with a well-ventilated root zone and irrigation water not exceeding 1500$_{\mu}$mhos (1000 ppm).

Genetic diversity

Because of their dioecism and scattered distribution, date palms are inherently highly diversified. Numerous cultivars, including dwarf and seedless forms occur, for example 394 in Libya, 352 in Iraq. Breeding programmes have been impeded by inability to select the male varieties for fruit characters (Barrett 1973). The most immediate improvement will come from direct introduction of cultivars, selection for local adaptation and the development of rapid micropropagation techniques.

Production season and time of cropping

The production season in Australia is from April through June, depending on variety. Late varieties would require low humidities during June, and artificial ripening of fruit.

Yield potential

For well-maintained orchards the yield potential would be: Africa 35 kg per palm; North America 120 kg per palm; Australia (experimental) 90 to 120 kg per palm (Freeman 1936–37). Commercial cropping commences in year 6 and continues until the fruit becomes inaccessible, usually around year 50.

Market acceptability

Most dates are pressed into solid blocks for marketing. Australian imports exceed 4 Kt. Top quality fruit of semi-dry varieties such as Deglet Noor may be marketed still attached to the spikelet and packed in individual boxes.

Export potential

World trade is 2 to 3 Mt annually and demand seems likely to overtake supply. Freight costs per dollar product should be low.

Cultural details

See Nixon (1969) and Freeman (1936) for comprehensive treatments.

Propagation

- Seed — only for selection areas (could be used as wind-breaks). Seed from shop fruit is often dead because of steam sterilization procedures.
- Suckers are at present the only form of vegetative propagation. They are allowed to develop for several years and have their own roots when removed. It is sometimes necessary to mound soil around the stem to induce root development. Special tools have been designed for this work (Freeman 1936).
- Successful tissue culture has been reported (Tisserat 1979).

Management

Operations specific to this crop include:

- Removal of unwanted suckers and those fronds which have become overshadowed.
- Collection of pollen and artifical pollination (Burkner and Perkins 1975).
- Bunch pruning to increase fruit size.
- Placement and protection of the developing bunches (Miremadi 1971).
- Bunch harvesting and fumigation.
- Fruit grading and cold storage (also post-harvest ripening for some varieties).

Methods of pollination range from placing freshly opened male clusters over the top of the female spathe, to air blasting with pollen collected from special male orchards. One male is required for 25 females.

Irrigation

Similar to that for citrus for best yields.

Fertilization

- Organic manure, preferably ploughed in after a leguminous green crop.
- 10 kg per mature palm of a 1:3:3 (N:P:K) mix (Freeman 1936).

Pests and diseases

There are no major parasites in the field. Ripe fruits are attacked by a range of storage pests and consequently fumigation is the first post-harvest operation.

Harvesting methods

In developed countries all varieties are bunch harvested and the fruit mechanically separated from the bunch in the field. Unripe fruits are placed in ripening chambers followed by dehydration or rehydration if necesary.

Marketing characteristics

As the fruit is a stable product, it is readily transported. The value per weight ratio is high.

Storage characteristics

- Dry storage in sealed containers.
- Cold storage of graded fruit. This reduces pest build-up and retards conversion of sucrose into the more hygroscopic glucose and fructose.
- Long storage life.

References

Barrett, H.C. (1973), Date breeding and improvement in North America. *Fruit Var. J.* **27**, 50.

Burkner, P.F. and Perkins, R.M. (1975), Mechanical extraction of date pollen. *Date Growers Institute* **52**, 3.

Freeman, H.J. (1936–37), Date culture in Queensland, *Queensland Agricultural Journal* **45**, 376; 487; **46**, 392.

Miremadi, A. (1971), Principles of date pruning in relation to fruit thinning. *Date Growers Institute* **48**, 9.

Nixon, R.S. (1969), Growing dates in the United States, Agricultural Information Bulletin 207, USDA.

Tisserat, B. (1979), Propagation of date palm *(Phoenix dactylifera* L.*)* in vitro. *Journal of Experimental Botany* **30**, 1275.

Further reading

Barger, W.R. (1940), Handling and storing small lots of dates at home. USDA Circular 553.

de Fontenay, U. and V.J. (1960), Date growing in Australia. *Journal of the Australian Institute of Agricultural Science* **26**, 246.

subfamily Cocoinae

PEACH PALM (PEJIBAYE, PUPYNHA) by D.H. MAGGS
Bactris gasipaes HBK
Guilielma gasipaes HBK L.H. Bailey; *G. speciosa* Mart.;
G. utilis Oerst.; *G. chontadura* Triana

Botany

Peach palm is in the same subfamily as coconut and oil palm. *Bactris gasipaes* is the only species in the genus.

The species is a feather palm, forming a clump of stems by suckering from the base. It grows up to 20 m high and 15 cm in diameter. Stems are usually heavily armed with rings of very sharp black, 5 cm spines; the outer layer is very hard. Leaves are pinnate, 2.5 m long; leaflets are dark green, smooth, 50 cm long; leaves are deciduous leaving the trunk clean.

Flowers are in spikes produced in axils of the lower leaves; monoecious, the females are interspersed among the more numerous males. Inflorescences develop into pendant bunches (up to 12 kg) of fruits (up to 300). Fruits are with kernel, shell (endocarp), pulp (mesocarp), skin (pericarp), 2 to 6 cm long; mesocarp orange, rich in vitamin A, nicotinic acid, very starchy.

Fruit composition (part-estimated from related species)

	Percentage of fruit	Percentage of part: Water	Protein	Fat/oil	Carbohydrate
Pulp	60	50	3	7	40
Shell	30	—	—	—	—
Kernel	10	10	8	50	20

Peach palm is widely distributed throughout tropical America from Colombia to Costa Rica. It is not known as a wild species in Costa Rica where it is most cultivated. Some of the uses of peach palm are:

- Fruits eaten after boiling 3 hours in salt water.
- Staple diet of local population for several months.
- Canned; halves used with dips at parties; terminal bud of over-tall stems eaten as palm cabbage. (This does not destroy the clump as would be the case with single-stemmed palms.)
- Palm wine.
- Outer wood used for carving or where very hard material is required.

Climatic requirements

Peach palm requires a strictly tropical climate with a temperature range of 16 to 25°C, rainfall up to 2500 mm annually, and an altitude between sea level and 1200 m.

Soil requirements

Peach palm grows on heavy soils.

Genetic diversity

The species is known to have been in cultivation for more than 500 years and must have been subjected to considerable selection. Numerous varieties are reported but not described. There are variations in pericarp colour, fruit size, flesh thickness and texture. Seedlessness and spinelessness are in part heritable. Peach palm could possibly be hybridized with *Bactris* spp.

Production season and time of cropping

In Costa Rica (10°N) the palm flowers at the end of the dry season (March). It ripens from September on, throughout the wet season. Fruit will hang until March, but it is difficult on the older tall stems to harvest individual ripe fruits.

Yield potential

For a well-grown clump the yield potential is:

maximum individual bunch weight	12 kg
fruit (80% bunch weight)	9.6 kg
bunches per stem per year	3 (up to 13 reported)
stems per clump	4
total annual production	115 kg
equivalent orchard yield	23 tonnes per hectare

Market acceptability

Its present use in Central America is as a staple food equivalent to a yam or potato. It is sold in the streets as a fast food but there must be reservations as to its acceptance in Australia.

Export potential

Export potential of the fruit would be as a canned speciality.

Cultural details

Propagation is by suckers from a mature palm after it has produced four stems. Suckers should have produced roots and be 7 to 8 cm in diameter at the base and 1 to 1.5 m high. The foliage should be cut back and the plant shaded until established. Seedlings fruit at 8 years old. Removal of spines causes excessive bleeding and is said to weaken the palms.

Harvesting methods

Harvesting is done manually or by gravity. Shaking is also possible.

Marketing characteristics

The fresh fruit has several weeks' shelf life and this could probably be increased by appropriate technology. The processed fruit is sold ready to eat.

Storage characteristics

Fruit last several months on the plant. It can be dried after processing and stored for up to 6 months.

Further reading

U.S. Nat. Acad. Science (1975), Under-exploited tropical fruits with promising economic value. Pejibaye, p. 73.

Uphof, J.C.T. (1968), Dictionary of economic plants. Cramer, Lehre, Holland.

Johanssen, C.K. (1966), Pejibayes in commercial production. *Turrialba* **16,** 181.

Popenoe, W. and Jiminez, Oton. (1921), The Pejibaye, a neglected food plant of tropical America. *Journal of Heredity* **12,** 154.

COCONUT PALM

by M.A. FOALE

Cocos nucifera L.

> ... and from these trees and their fruit are made the following things: sugar, honey, oil, wine, vinegar, charcoal, and cordage ... and matting ... and it serves them for everything they need. And the aforesaid fruit, in addition to what is thus made of it, is their chief food, particularly at sea.
>
> Manuel, King of Portugal, 1501

There is no evidence that coconut palms grew in Australia before European settlement. They were probably established at many points on the north coast between Broome and Rockhampton during the 19th century. This is one of the major food crops of well-watered tropical areas of low altitude and fully supports human life on atolls, in combination with fish and very limited green and starchy vegetable supplies.

Botany

Cocos nucifera is a monoecious palm that carries 20 to 40 fronds 3 to 6 m long, producing 8 to 20 new fronds per year. Each frond has an axillary bunch of 0 to 20 nuts (in exceptional circumstances more). The bunch bears 1 to 3 female flowers (2 cm in diameter) per branch on the lower part of the lower branches, totalling up to 50 flowers per bunch, and thousands of 3 mm male flowers in the distal ends of all branches. Pollen is shed before female receptivity, forcing cross-pollination in most cultivars, but certain dwarf cultivars have coincidence of the male and female phases, resulting in a variable proportion of selfing.

Nut development requires approximately 12 months; germination begins 1 to 4 months from nut maturity, and the first flower appears 3 to 10 years after germination, depending on cultivar and conditions. Vigorous palms produce 80 to 180 nuts per annum.

Palms can reach a height of 25 m in 80 years, but rarely live beyond 60 years because of increasing trunk fragility.

Climatic requirements

The coconut thrives at sea level to 300 m within 10° of latitude of the equator in a diurnal temperature range of 22° to 32°C with scant annual fluctuation. As both the diurnal and annual ranges expand with increasing distance from the tropics, palm growth rate and productivity are reduced. The geographical limit for reproduction is about 23° of latitude but palms grow and flower fruitlessly at 27°, for example in Brisbane. Rainfall exceeding 2000 mm is required though less suffices if there is root access to ground water, or if the temporal distribution is uniform.

Soil requirements

It is a very adaptable species that grows on coarse coralline and volcanic gravels right across the textural range to clay loams over coral and heavy marine clays with artificial drainage. It is not adapted to waterlogging or high salinity but is tolerant of salinity where tidal influence on underground water ensures regular flushing of fresh water through the root system. Coconut soils range in pH from 3.4 to 9. Water must be readily available at all times to achieve potential growth as, in common with most palms, there is no dormant growth phase.

Table 1. Some coconut cultivars

Variety (1)	Trunk type (2)	Frond length (m)	Crown type	Best spacing (m)	Fruit size	Predominant fruit colour	Fresh weight of meat (g)	Time to flower (yr)	Breeding system	Genetic variability
Tall — Asian	T.R.W.	4.5–6	Semi-compact	8	med/large	green	300–450	6–8	outcross	high
Tall — Pacific	T.R.W.	4.5–6	lax	8	med	bronze	280–350	5–7	outcross	high
Dwarf — Malayan	N.F.N.	3–4.5	lax	6.5	small/med	orange & yellow	250–300	3–5	selfing (80%)	low
Dwarf — Fijian	T.I.N.	3–5	compact	6.5	med	green	280–350	5–7	outcross	intermediate

(1) Most coconut palms in the Northern Territory and Western Australia appear to be of Asian origin, while those on the Queensland coast are mostly of Pacific origin. In their general appearance these types are hard to distinguish except on fruit colour and size.

(2) 1st letter: T = tapering from thick end base; N = narrow base.
2nd letter: R = robust trunk up to 60 cm diam. in lower part; F = fragile trunk 25–40 cm diam.; I = intermediate.
3rd letter: W = wide space between leaf scars on trunk; N = narrow spacing between scars.

Genetic diversity

There are many cultivars and variants especially in the west Melanesian region of the Pacific which is the most likely centre of origin. Two broad groups are recognized and are known as Talls and Dwarfs. Talls have longer fronds, more robust trunks, are slower to begin flowering (5 to 10 years), and are predominantly out-pollinating. Dwarfs are smaller in all physical attributes and as they are predominantly self-pollinating there is considerable uniformity and several recessive colour genes are expressed in the colour of the petiole and the immature fruit. The common Dwarf colours, apricot, orange and yellow are rarely seen in Talls. Dwarfs are less adaptable and less productive per palm than Talls, but they are resistant to several serious virus (or mycoplasma) diseases that are fatal in many Tall cultivars.

Hybrids between certain Talls and self-pollinating Dwarfs are highly productive and much in demand for new plantations.

There is apparently considerable genetic adaptation to local rainfall and soil environments, and also seasonally lower temperatures, for example, in the Loyalty Islands, Tonga and the southern Cook Islands (Latitude 21°S). Resistance to insect pests and fungal diseases varies considerably between cultivars.

Production season and time of cropping

Production is non-seasonal in coastal areas within 15° of the tropics. Low temperatures (minimum below 20°C) and dry seasons slow the growth rate but there is no true break in seed production.

Yield potential

Nut size is genetically controlled and highly variable, especially in the heterozygous Tall group, while nut number varies with conditions. Yield expressed only as nut number is consequently of limited use. The products of most trade interest worldwide are dried endosperm (copra), oil, which is extracted from copra (66%), and dried grated endosperm (desiccated coconut). Seeds for planting, de-husked nuts for culinary or novelty trade, moist endosperm (meat) for culinary use, whole nuts (complete with exocarp or husk) for ornament production are items on local markets only. Fibre is a by-product extracted from husks in some countries and is used for rope, mat, and mattress production. It has potential as a filler in horticulture potting mixtures.

Table 2. Annual yield of various coconut products

	Per palm	Per hectare (180 palms)
Nuts	50 – 150 kg	9020 – 27000
Copra (6% moisture)	10 – 30 kg	1800 – 5410
Oil	6.6 – 20 kg	1190 – 3600
Endosperm (50% moisture)	18 – 54 kg	3250 – 9740
Fibre	5 – 15 kg	900 – 2700

Marketing characteristics

A trade in de-husked nuts is currently supplied solely with imported nuts from Tonga, Fiji, New Guinea, Malaysia and the Philippines. Local suppliers could enter this market. The Brisbane market, which is most active in the cooler months, consumes 20 000 nuts per month.

There is a strong demand on the north coast, especially Darwin, Broome, and Cairns, for coconut seedlings. The more attractively coloured Dwarfs command a premium. Demand currently exceeds supply of seedlings. Ten thousand palms have been planted in Darwin since the 1974 cyclone.

Export potential

The prospects for an export market are limited.

Cultural details

Nursery and transplanting management is more demanding in northern Australia than in the Pacific Islands due to the severe dry season and the generally poor soils. The large endosperm gives seedlings great resilience up to 9 months of age against transplanting shock, but poly-bag seedlings give best results at all ages. Spacing at 6 to 8 m isometric is preferred.

Irrigation is needed for the period that it takes root systems to reach the dry season water-table. If there is no water-table, or if the soil texture prevents deep rooting, permanent trickle irrigation is needed.

N, P and K all are commonly needed in northern Australian soils. On sandy soils K fertilizer rate should be high. N supply should be conservative in the nursery to reduce leaf spot susceptability. Twice yearly subsurface probe dressings of fertilizer (2:2:4), at 100 g per tree each time in the first year, should rise steadily to 500 g each dressing in year 5 and be varied thereafter according to growth indications.

Pests

Two green caterpillar species, *Cephrenes trichopepela* and *C. orangiades,* severely reduce leaf area on seedlings and small transplants. Grasshoppers are a sporadic problem while Darwin has a borer, *Diocalandra frumenti,* that attacks frond petioles and female flowers. *Aspidiotus* leaf scale has been noticed in several locations, but is not serious. *Brontispa longissima* occurs unobtrusively on the north Queensland coast where it was positively identified for the first time in 1976. It appeared for the first time recently in Darwin where it causes severe leaf damage. *Mastotermes* spp. often damage trunk bases, sometimes fatally.

Diseases

Drechslera incurvata is sometimes serious in nurseries but responds to fungicide sprays. Bud rot is causing increasing concern in the Darwin area.

Harvesting methods

In pilfer-free, high humidity situations natural fall of ripe nuts removes the need to hand harvest from the palms. In northern Australia, especially in the dry season, nuts are severely desiccated on the bunch and must be harvested to be of use. This is rarely practised, so serious wastage occurs. A hook knife on a light pole can be used satisfactorily to harvest whole bunches to heights of 15 m.

Marketing characteristics

The coconut seedling trade is well developed locally in centres like Darwin, Broome and Cairns.

De-husked nut marketing is well established. It will probably be many years before local underdeveloped communities (by European-Australian standards) such as Bathurst Island have the motivation to process nuts for the de-husked trade. Current wage rates are prohibitive and the processing skills are not developed.

Storage characteristics

De-husked nuts must be protected from very low humidity and high temperatures to avoid cracking. Shelf life is 2 to 3 months, ending when the liquid endosperm (coconut water) has evaporated or when the shell cracks due to desiccation.

Transport characteristics

There are no special problems with transporting nuts. De-husked nuts are carried in bags or cartons.

Processing characteristics

Local oil production from sun-dried copra is a possibility in isolated coastal communities that currently import cooking oil. Appropriate hand-operated presses are available, for example, in India. De-husking skills have not yet been developed.

Acknowledgements

For local information included here thanks are due to Bruce Toohill (West Australian Department of the North West), Leon Radunz and Rex Pitketchley (Northern Territory Department of Primary Production), George Brown (Darwin City Council), Les Brigden (CSIRO Division of Forestry), and Brian Watson (Queensland Department of Primary Industries).

Further reading

Foale, M.A. (1968). The growth of coconut seedlings. *Oleagineux* **23**, 651.

Williams, C.N. (1975). Coconuts *in* 'The agronomy of the major tropical crops'. Oxford University Press, London.

Chapter 21

PASSIFLORACEAE

PASSIONFRUIT
Passiflora spp.

by P.R. BEAL and P.J. FARLOW

Botany

The family Passifloraceae includes 550 species in 12 genera (Killip 1938). The most important genus *Passiflora* has about 400 species which are mostly indigenous to tropical and South America but include about 40 in Asia and the South Pacific Islands. *Passiflora* species have easily recognized bisexual flowers generally with a colourful and complex corona and a superior ovary on a gynophore. They have tendrils to aid their climbing habit and their alternate leaves often have glandulate petioles. The fruit is a capsule or berry containing numerous seeds with a pitted testa surrounded by a fleshy aril.

The ornamental flowers and edible fruit of many species have influenced their spread outside their original areas of distribution. Although 50 or 60 species bear edible fruit (Martin and Nakasone 1970) most of these are unknown outside of the limited areas where they grow wild or are sometimes cultivated.

Climatic requirements

Passionfruit are adapted to tropical and subtropical areas with a moderately high rainfall. The purple passionfruit *Passiflora edulis,* which is native to Brazil, thrives in the warm humid coastal areas of eastern Australia and on the tropical highlands. Although it withstands light frosts it is injured by temperatures 1 to 2°C below freezing and commercial sites should be free of frost. The golden passionfruit *(P. edulis* f. *flavicarpa),* the granadilla *(P. quadrangularis)* and water lemon *(P. laurifolia),* in particular, are less cold tolerant and require more tropical conditions than *P. edulis.* However, some species are more cold tolerant than *P. edulis,* such as *P. incarnata* (indigenous to south-east United States), *P. ligularis* (indigenous to mountainous areas of Mexico and Central America) and species in the subgenus *Tacsonia* such as *P. mollissima* and *P. antioquiensis* (which are adapted to high altitudes). Pattersen *et al.* (1976) classified species in the subgenus *Granadilla* in order of increasing chilling tolerance as *P. edulis* f. *flavicarpa, P. maliformis, P. incarnata, P. edulis* and *P. caerulea.*

Soil requirements

Experience with *P. edulis, P. edulis* f. *flavicarpa* and *P. quadrangularis* in Queensland indicates passionfruit are not generally exacting in their soil requirements. They do best on fertile well-drained loams. The main needs are good soil moisture combined with adequate subsurface drainage. Vines succumb rapidly to waterlogging and on shallow soils are best planted on raised beds. They grow best in soils with pH above 6.0, and on strongly acid soils (below pH 5.5) lime is required.

Genetic diversity

Only about 40 to 50 species have been investigated agronomically or cytologically by workers in various countries. These mainly comprise edible types in the subgenera *Granadilla* and *Tacsonia,* weed species, the native Australian species as well as solely ornamental types. Also, little breeding has been carried out and the genetics of the genus is poorly understood. However, substantial diversity, which is potentially most useful in passionfruit improvement, is known to exist within and between species for pollination system, vigour, yield and production season, climatic and ecological adaptability, tolerance of passionfruit woodiness virus (PWV), resistance to fusarium wilt (Purss 1958), resistance to phytophthora blight, brown spot, septoria spot, nematodes and fruit characteristics including size, pulp content, colour and flavour.

Storey (1950) and Beal (1968) have now established chromosome numbers in 46 species or forms in the genus. The numbers 2 n = 12, 18, 20, 24, 36 were determined although other numbers of series could exist considering the large number of uncounted species. The 2 n = 18 group species (including the subgenera *Granadilla* and *Tacsonia*) generally have vigorous growth and show the greatest expression of flower and fruit development. The 2 n = 12, 20, 24 species tend to have small fruit with insipid taste and vines of moderate vigour.

Breeding and selection for improved agronomic characteristics in the major edible species and evaluation of rootstocks have been emphasized in Australia and Hawaii in the last 20 years. Selection has been practised for improved tolerance of alternaria spot and resistance to fusarium wilt in *P. edulis* f. *flavicarpa* in Hawaii (Ito 1978), and Australia respectively. *P. edulis* x *P. edulis* f. *flavicarpa* hybrids combining high production, winter and summer cropping, tolerance to PWV, nematodes and fusarium wilt were selected by Groszmann and Meurant (personal communication) in the early 1960s and have been exploited in Queensland since that time. Interspecific hybrids involving species within the subgenus *Granadilla* were produced by Beal (1972) and Ruberte-Torres and Martin (1974) some of which were fertile. This is contrary to the contention that interspecific hybrids were invariably sterile in spite of ease of crossing (Storey 1950).

Production season and time of cropping

Production season in passionfruit is strongly influenced by the nature of the species exploited although cultural procedures and location are modifying factors.

The purple passionfruit produces a heavy summer crop followed by little or no winter crop in Queensland. The golden passionfruit differs from the purple passionfruit in having a single extended production season over late summer-autumn and early winter. Cropping of commercial passionfruit hybrids in Queensland is continuous with peaks in summer and winter.

The variation in time of cropping known in *P. edulis* and its intraspecific hybrids has also been observed in Queensland between species such as *P. incarnata* and *P. herbertiana* (autumn cropping), *P. seemanni* (winter), *P. quadrangularis* (autumn and spring), *P. mollissima* and *P. antioquiensis* (spring and summer cropping).

Yield potential

The purple passionfruit generally yields about 5 to 10 tonnes per ha per year when PWV is not severe while commercial *P. edulis* hybrids yield up to 25 tonnes per ha per year. In contrast yields of 35 to 45 tonnes per ha per year have been recorded in commercial plantations of the golden passionfruit in Hawaii (Anon. 1954) and in South Africa, and in *P. mollissima* in Colombia (Martin and Nakasone 1970). Yields of 70 to 150 kg fruit per vine per year are produced from *P. quadrangularis*.

The average yields in Queensland and in New South Wales are around 14.8 and 11.8 tonnes per ha per year respectively. More favourable growing conditions in Queensland account for the higher yields recorded in this State. The golden passionfruit does not always yield as well as its profuse flowering would indicate. Self-incompatibility is known in *P. edulis* f. *flavicarpa*, *P. cincinnata* and *P. laurifolia* and the need for hand pollination is recorded in *P. quadrangularis*. *P. laurifolia* grows well and yields successfully in north Queensland only.

The influence of season in reducing pulp content in the *P. edulis* hybrids 3–1 and E23 in March–April and October–November was recorded by Bradley (1972).

Market acceptability

Market acceptability is related to traditional consumer preferences and whether a fresh or processing market is being supplied. *P. edulis* hybrids, as well as *P. edulis* f. *flavicarpa* and *P. ligularis* are the most common passionfruit of Australia, Hawaii and Mexico respectively. The fresh market in Australia demands a blemish-free, well-filled, sweet, flavoursome fruit with an attractive purple-black skin. This requirement is currently met by the *P. edulis* hybrid Purple Gold. This preference by Australian consumers is probably related to past exploitation of the purple passionfruit which is not now commercially important.

The processing market is more discriminating than the fresh market with respect to pulp content and flavour and to a lesser extent acidity and SS.

Export potential

Australian passionfruit production fluctuated from 3654 to 2966 tonnes each year from 1972 to 1978. Imports of juice plus pulp in the same period ranged from 139 to 1039 tonnes fruit equivalent. Local processors have claimed that the Australian market is currently completely satisfied with 2500 tonnes of fresh product intake. The overall market for passionfruit products may require further market development, product innovation and possibly some penetration into overseas markets. Europe and the United States may offer some scope for expansion. However transport costs are a major problem facing exporters.

The high cost of processed passionfruit relative to other fruit juices is a major factor restricting expansion.

In other countries commercial passionfruit production is based on lines of *P. edulis* f. *flavicarpa*. Countries involved include Hawaii, Brazil, Sri Lanka, Fiji, Papua New Guinea, West Samoa, France, Malaysia, Cook Islands, Nuie and African countries. Many of these countries can produce low cost pulp because of the availability of cheap labour. Recently soil borne diseases have reduced production in Brazil and Sri Lanka. Some of these countries experience fruit setting problems because of the self-incompatibility of the *flavicarpa* material they are using.

Cultural details

Propagation

Species of *Passiflora* are generally readily propagated by seed, cuttings and air layers. In Queensland commercial *P. edulis* hybrids, which do not come true to type from seed, are grafted on to the stock *P. edulis* f. *flavicarpa* which is resistant to fusarium wilt and nematodes. The present authors have also found that many species may be readily grafted onto *P. edulis* f. *flavicarpa* and grow satisfactorily giving a quick and acceptable method of propagation.

Management

Successful passionfruit growing is dependent on selection of suitable location, good land preparation, and management expertise in trellising, pruning and training, irrigation, nutrition, pollination needs, pest and disease control, weed control and harvesting. In Queensland about 540 to 790 vines per hectare are established for *P. edulis* hybrids.

Such cultural requirements for commercial production of *P. edulis* and its hybrids, *P. edulis* f. *flavicarpa* and *P. quadrangularis,* have been well described in various countries including Australia, Hawaii and South Africa.

Irrigation

Adequate soil moisture during flowering and fruit development is necessary to obtain maximum yields. Some form of irrigation (overhead or trickle) is necessary when rainfall is normally low or unreliable. Lack of soil moisture in spring and early summer can directly reduce the summer crop and also affect the development of flowering laterals.

Fertilization

Passionfruit require a plentiful supply of plant foods to maintain growth and yield. In some situations small quantities of a complete chemical fertilizer are applied several weeks before planting. Subsequent applications are made during the year, and timing, quantity, and type of fertilizer used depends on the cropping potential of the vines, irrigation practices and rainfall. During winter foliar sprays may be used. In Queensland rates of 0.8 to 1.2 kg N, 0.3 to 0.5 kg P and 1.1 to 1.5 kg K are applied per vine per year.

Pests and diseases

Major problems in *P. edulis* and/or *P. edulis* f. *flavicarpa* are PWV (a necrotic strain of this virus [PWV-N] being a major problem in Queensland at present), base rot (undetermined), septoria spot *(Septoria passiflorae),* phytophthora blight *(Phytophthora nicotianae* var. *parasitica)* and brown spot *(Alternaria passiflorae).*

Other problems which can be severe at times are fruit fly, mites, scale insects and cucumber mosaic virus (CMV). CMV is a problem in cooler areas (for example New South Wales) The synergistic action of mutual infection with CMV and PWV causes rapid death of vines. Fusarium wilt *(Fusarium oxysporum* f. var. *passiflorae)* and nematodes *(Meloidogyne javanica)* are not problems in Queensland at this time.

Brown spot was a severe problem in *P. edulis* but has only been a major problem in the hybrids. In contrast, septoria spot which was not of substantial concern in *P. edulis* is a major problem in the hybrids. The hybrids are susceptible to leaf and fruit infections of phytophthora blight which favours warm wet conditions. However, the use of resistant *P. edulis* f. *flavicarpa* rootstock has provided effective protection against phytophthora infections of root and stem.

Harvesting methods

Orchard trellis or arbor systems are planned to minimize high harvesting costs. Fruit of *P. quadrangularis* is clipped in commercial harvests and handled very carefully thereafter. Fruit of *P. edulis* hybrids and *P. edulis* f. *flavicarpa* are normally allowed to fall on the ground and collected once or twice per week in Hawaii (Anon. 1954) and daily or more frequently in Queensland to reduce sunburn. Also, it is recommended that fruit be kept as cool as possible during handling (Willis, Stephens and Groszmann 1961). A prototype mechanical harvester has been developed in Queensland to overcome the costs of manual harvesting.

Marketing characteristics

In Australia, approximately 3000 tonnes of passionfruit are produced annually at a value of about $1.8 million. More than 80% of Australia's passionfruit is produced in south-east Queensland. Queensland's production of 2780 tonnes in 1979 was consigned to the local fresh market (7%), interstate (25%), and to processors (68%).

Import of pulp and juice (see section on export potential) mainly from Fiji and Sri Lanka increased the total Australian apparent consumption.

The principal processed passionfruit products are juice and pulp. The major uses for these are as ingredients in yoghurts, ice cream, jams, flavouring and in fruit juices.

Storage characteristics

Passionfruit stored at room temperature for a few days will shrivel as the rind dries. Fruit will store for up to 4 to 5 weeks with minimum loss of fruit weight if rots are controlled and if held at 5°C and 80 to 90% relative humidity. Waxes and fungicidal dips are often used to extend shelf life in fresh fruit and prevent withering of skin and rots.

Passionfruit pulp and juice are normally extracted, sterilized by heat and held in frozen storage (–18°C) in 20 kg containers. Manufacturers generally specify microbial counts before final use.

Transport characteristics

In Australia fruit of *P. edulis* hybrids are transported by road and rail to the fresh market in 10 kg cartons and to the processor in bags or bulk bins. The fruit is well protected by the tough leathery rind and is very tolerant of the rigours of transport. The species *P. ligularis* and *P. maliforms* have much harder rind than *P. edulis*, whereas *P. quadrangularis* is soft skinned and sensitive to damage. The last species is generally packed and consigned in crates lined with wood wool.

The holding conditions for pulp and juice which would be required during transport are described in the section on storage characteristics.

Processing characteristics

Australian processors have generally demanded that fruit for processing have the same flavour as *P. edulis* whereas the more acid, aromatic and orange pulp of the golden passionfruit is traditionally processed in Hawaii. Bradley and Jordan (1968) showed that the old commercial hybrids Redlands Triangular (3–1) and Redlands Pink (3–26) were as good as *P. edulis* for the manufacture of carbonated juice but that *P. edulis* f. *flavicarpa* was not suitable. The lower ester value (83 ppm) recorded for *P. edulis* f. *flavicarpa* compared with the hybrids (122 and 159 ppm) suggested that this may have influenced panel unacceptability. Chemical identification of the volatile flavouring constituents involved may be useful to test this suggestion.

The hybrid E23 is the highest yielding commercial cultivar and is readily acceptable by Australian processors because of its desirable flavour. Also, E23 has a pulp content ranging from 40 to 50%, is less subject to seasonal variation than the hybrid 3–1 (35–50%) and is less likely to fall below the 38% minimum acceptable level (Bradley 1972).

The SS and pH are of less importance, as passionfruit juice is used primarily in low concentration for flavouring, and sugar and citric acid levels can be easily altered.

References

Anon. (1954), Passionfruit culture. University of Hawaii Extension Circular 345.

Beal, P.R. (1968a), Queensland Department of Primary Industries Report.

Beal, P.R. (1968b), The cytology of the native, naturalised and some recently introduced species of *Passiflora* L. University of Queensland thesis.

Beal, P.R. (1972), Two new interspecific hybrids in the genus *Passiflora*. *Sabrao Newsletter* **4,** 113.

Bradley, B. (1972), Queensland Department of Primary Industries Report.

Bradley, B. and Jordan, R. (1968), Queensland Department of Primary Industries Report.

Ito, P.J. (1978), Noels Special Passionfruit. *HortScience* **13,** 197.

Killip, E.P. (1938), The American species *Passifloraceae*. Field News, Nat. Hist. Bot. Series 19 parts 1 and 2 publ. 407 and 408, Chicago.
Martin, F.W. and Nakasone, H.Y. (1970), The edible species of *Passiflora*. *Economic Botany* **24**, 333.
Patterson, B.D., Murata, T. and Graham, D. (1976), Electrotype leakage induced by chilling in *Passiflora* species tolerant to different climates. *Aust. J. Plant Physiol.* **3**, 435.
Purss, G.S. (1958), Studies of the resistance of species of Passiflora to fusarium wilt *(F. oxysporum* f. *passiflorae). Queensland Journal of Agricultural and Animal Sciences* **15**, 95.
Ruberte-Torres, R. and Martin, F.W. (1974), First-generation hybrids of edible passionfruit species. *Euphytica 23*, 61.
Storey, W.B. (1950), Chromosome numbers of some species of *Passiflora* occurring in Hawaii. *Pacific Sc.* **4**, 37.
Wills, J., Stephens, S.E. and Groszmann, H.M. (1961), Growing passionfruits and granadillas. *Queensland Agricultural Journal* **87**, 680.

Table 1. Climatic tolerance of selected *Passiflora* species

	Native distribution (Killip 1938)	Location	
		°Latitude	Altitude m
Tropical:			
P. laurifolia	West Indies to north of South America	0 – 10	—
P. quadrangularis	Colombia	0	1100
P. maliformis	Colombia	0	1200
P. edulis f. *flavicarpa*	—	—	—
Temperate:			
P. edulis	Brazil to Argentina	0 to 30	—
P. incarnata	South-east U.S.A.	30 to 40	—
P. caerulea	Brazil to Argentina	30	—
P. ligularis	Mexico to north of South America	10	1000
*P. mollissima**	Andes	20	1000 to 2000
*P. jamesonii**	Ecuador	0	4000

* subgenus *Tacsonia*

Table 2. Pulp colour in species of *Passiflora*

Black	*P. penduliflora, P. suberosa*
Red	*P. gracilis, P. caerulea*
Orange	*P. warmingii, P. mollissima, P. edulis* f. *flavicarpa*
Yellow	*P. edulis*
Grey-yellow	*P. maliformis, P quadrangularis*
Grey	*P. aurantia, P. lingularis, P. incarnata, P. foetida*

(Beal 1968a)

Table 3. Pulp characteristics in *P. edulis, P. edulis* f. *flavicarpa*, hybrids 3–1 and 3–26 and in 36 other hybrid lines from the Redlands Horticultural Research Station (1967)

Accession	% Pulp	pH	%Av SS	% Av tit. acidity (citric)	Total volatile esters ppm (hexyl caproate)
P. edulis	49	3.1	15.3	2.4	159
P. edulis f. *flavicarpa*	37	2.9	15.3	3.4	83
3–1	46	2.9	14.5	3.2	122
3–26	57	2.9	14.1	3.4	149
Other hybrids	35–68	2.7–3.1	13.4–16.1	2.3–3.6	49–200

(Bradley and Jordan 1968)

Table 4. Pulp characteristics in seven species of *Passiflora* (1967)

Species	pH	TSS (%)	Titre 0.1 N NaOH to 10 mL juice
P. edulis	2.8	17.7	48.1
P. edulis f. *flavicarpa*	3.2	17.6	66.7
P. maliformis	3.4	16	43.2
P. subpeltata	5.1	13.4	6.7
P. mollissima	3.2	11	31.4
P. herbertiana	3.8	10.5	20.0
P. suberosa	5.7	11.5	—

(Beal unpublished)

Table 5. Occurrence of bivalents and micronuclei in pollen mother cells in six *Passiflora* hybrids and species (1966–1968)

Accession	Mean no. bivalents/pmc	Percentage of pmc with micronuclei
P. caerulea-quadrangularis	7.1	38
P. alato-caerulea	3.8	29
P. caerulea-racemosa	9.0	64
P. edulis-incarnata	9.0	21
P. edulis-caerulea	6.5	10
P. filamentosa	1.2	69
*P. subpeltata**	9.0	0

* normal behaviour

(Beal unpublished)

Table 6. Fruit characteristics in selected *Passiflora* species (minimum of 10 fruit per sample in August 1966, Ormiston)

Species	Fruit weight (g)*	Pulp percentage*	Mean number seeds per fruit*
P. edulis	36 ± 5	48 ± 5	122 ± 57
P. subpeltata	17 ± 5	34 ± 18	158 ± 43
P. quadrangularis (small)	870 ± 170	20 ± 5	159 ± 48
P. seemanni	68 ± 13	30 ± 5	332 ± 76
P. mollissima	41 ± 8	54 ± 22	110 ± 21
P. edulis f. *flavicarpa*	72 ± 7	35 ± 3	50 ± 48
P. suberosa	0.9 ± 0.1	—	29 ± 6.1

* ± Standard deviation

(Beal unpublished)

Number of fruit of P. edulis and the hybrids 3-1 and 3-26 produced at two-weekly intervals

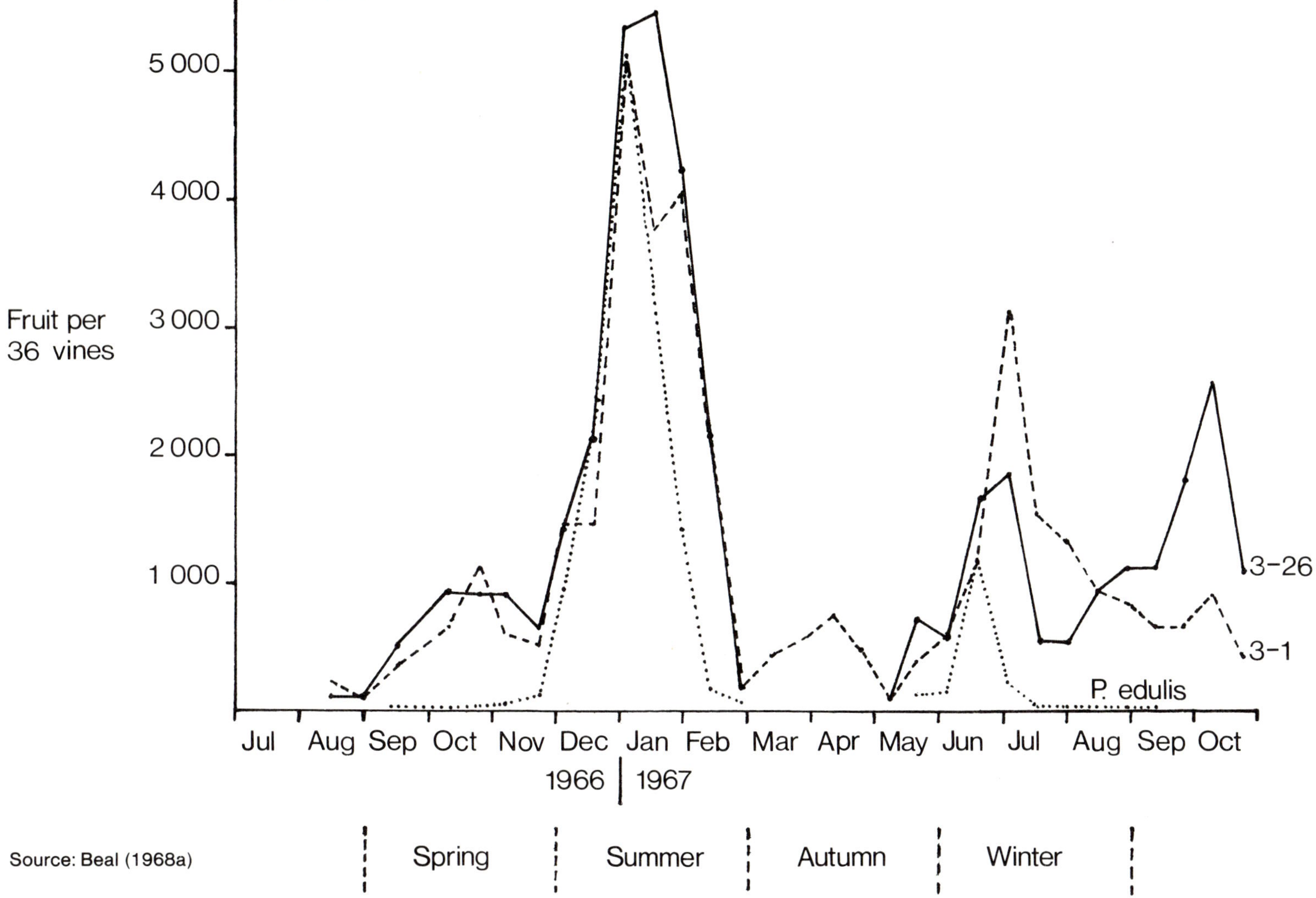

Source: Beal (1968a)

Distribution of Native Species of Passiflora in Australia

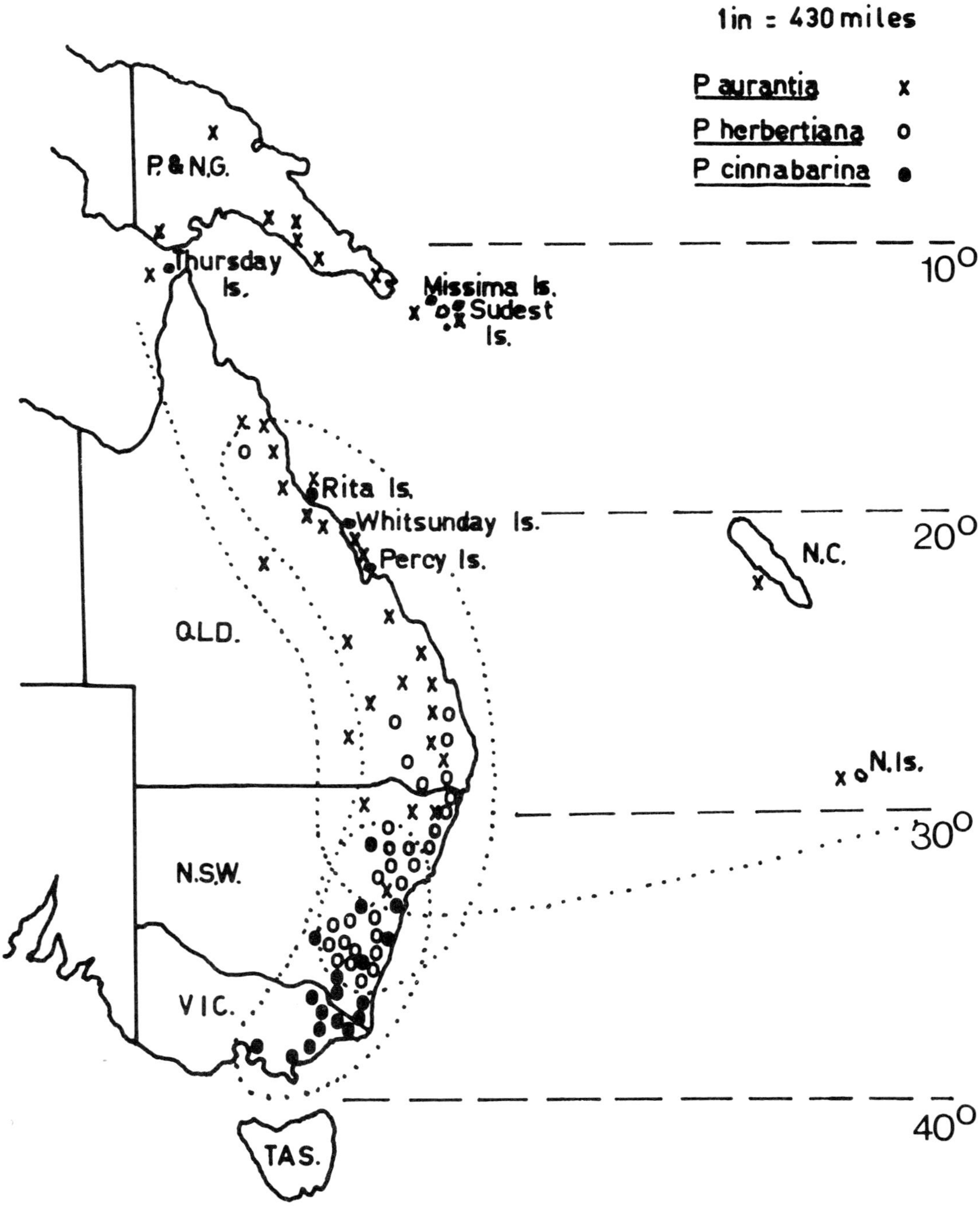

Source: Beal (1968b)

Chapter 22

PROTEACEAE

MACADAMIA

by B.W. CULL

Macadamia integrifolia Maiden & Betche
and **Macadamia tetraphylla** L.A.S. Johnson

Botany

The taxonomy of *Macadamia* has been confused almost from the first description by Dr Ferdinand von Mueller in 1857. Many revisions have been made and the presently accepted one is by Smith (1956) which is based on work by Johnson (1954).

The genus *Macadamia* belongs to the family Proteaceae, series Folliculares, and the tribe Grevilleae, which includes the well-known native ornamentals Grevillea, Hakea and Buckinghamia.

There are seven Australian species of *Macadamia* which fall into the following intrageneric groups:

M. integrifolia, M. tetraphylla, M. ternifolia;
M. whelani, M. hildebrandii;
M. praealta, M. heyana.

Three are recognized from New Caledonia:

M. roussellii, M. vieillardii, M. francii.

Of these species only *M. integrifolia* and *M. tetraphylla* are considered edible. Three other edible nuts, the gevuina nut of Chile, *Gevuina avellana,* the helicia nut, *Helicia diversifolia* and the rose nut, *Hicksbeachia pinnatifolia* belong to this family.

Two other species reported in the literature are *M. minor* and *M. lowii.* However, Smith (1956) classed these as synonyms of *M. ternifolia* and hence they have lapsed.

The two commercial species may be separated by the following descriptions:

M. integrifolia	*M. tetraphylla*
leaves usually in whorls of 3;	leaves usually in whorls of 4;
leaves 100 to 300 mm long;	leaves may grow to 500 mm;
adult leaves entire, few spines;	adult leaves always spiny;
petioles more than 5 mm;	petioles rarely exceeding 3 m; leaves often sessile;
new growth pale green;	new growth pink;
flowers creamy white;	flowers pink;
blossoms June to March or everbearing;	blossoms August to October producing one crop;
shell of nut smooth.	shell of nut pebbled.

The major botanical description to follow refers to *M. integrifolia* with additions for *M. tetraphylla* where the variation is significant.

Tree structure

The trees are large and spreading, attaining a height of 18 m and spread of 15 m. More upright types are present and are being selected because of their suitability for closer planting. The bark is rough (but not furrowed), brown, and dark red when cut. Branchlets are cylindrical and clothed with small raised lenticels.

Growth in the mature tree occurs in two flushes, a midsummer and spring flush. The latter is associated with, or follows flowering. In young trees four flushes may occur. Elongation may be 300 to 500 mm consisting of 7 to 10 nodes, and is defined at its base by a leafless node. This is formed from the rudimentary leaves of the tip of the previous flush, which do not expand with the new flush.

Each node bears three leaves *(M. tetraphylla,* four*)*, and in the axil of each leaf there are three buds arranged longitudinally. Usually only the top buds emerge, which may give branches with sharply acute angles with the trunk. Thus formed crotches can be weak and break out. The second bud will form a good angle and the third may be at a right angle. If the first bud moves away from a leafless bud then the angle will be wider and better. Lateral branching does not necessarily occur with each flush. If a growing point is removed the uppermost node will flush. If this node is leafy the shoots will be upright and one can be selected for a good leader. Where it is leafless the shoot may be angled and the tree may lose its symmetry.

The macadamia has proteoid roots. Malcolm and Trochoulias (1979) describe these as dense clusters of short lateral rootlets in well-defined rows around the parent root axis. The prime function of proteoid roots appears to be in increasing the surface area of the root system for maximum absorption. The stimulant for their formation is not known, but it has been found that soil transferred from well-established trees will promote the development of these roots in nursery plants. Nurserymen in California claim that vigour of seedlings is closely related to the degree of proteoid root development.

Flowers

Flowers are borne on long narrow racemes arising from the axils of leaves or the scars of fallen leaves. They may be borne on new growth if it is mature, but more often on the two, or three seasons' growth preceding the most recently matured flushes.

The raceme is pendant, from 100 to 300 mm long, with 100 to 300 flowers. As many as 500 have been recorded. Flowers are borne in groups of 2, 3 or 4, on pedicels 3 to 4 mm long in the axils of minute bracts arranged at regular intervals along the rachis of the raceme.

The flowers, about 12 mm long, are perfect but incomplete in that they have no petals, but four petaloid sepals. These form the perianth tube.

Within the perianth, in the centre of the flower, is the single carpellate, superior ovary. The ovary contains two ovules of which only one normally forms. When both form, 'twins' or half nuts are developed which are commercially unsuitable. Total length of style and ovary is about 7 mm and there is a very small stigmatic surface. Four perigynous stamens, each with two anther sacs about 2 mm long adhere about two-thirds of the way up the floral tube. The pollen is formed in two locules in each sac giving 16 in all. The anthers are curved over the tip of the style and the eight inner locules are pressed against the outer walls of the stylar knob.

The style begins to grow and bend 6 days before anthesis. After 3 days it pushes its central portion between the sepals. Two days later the anthers begin to discharge pollen on to the style. Near anthesis the sepals curl back and the anthers separate from the style leaving four clumps of pollen on the tip. All this time the style is growing and bending, and at anthesis it breaks through the sepals and springs free from the anthers and then proceeds to straighten.

Anthesis occurs during the day with a peak in the afternoon. Flowering is basipetal, proceeding down the raceme over a period of 1 to 5 days.

Self-pollination is obviously very active but Urata (1954) has shown that the style is not generally receptive for 24 hours after pollination. Pollen germination is delayed for this time. He also showed that varieties vary from being totally self-compatible to being almost self-sterile. Due to the potential loss of pollen by rain and the self-sterility factor, cross-pollination is important. Wind is believed to play only a minor part, and bees are considered to be the major agents. Cross-pollination by hand has been shown to increase nut set and nut quality.

It is difficult to establish the practical impact of these findings in the field. Multiple variety plantings on a row basis have been made together with the introduction of high populations of bees, but more as an insurance than a necessity.

Most flowers fall within 10 to 15 days after anthesis. The remainder, of which there are many, have fertilized and enlarged ovaries. A continual drop of this young fruit extends from 45 to 60 days after anthesis. The loss of nuts at this stage and its cause is a major concern of research aimed at increased productivity. Nut drop appears to cease by mid-December in Queensland. Loss of nuts after that time will be due to insect attack.

The mature fruit is a follicle, green, globose and slightly oblique, 25 mm or more in diameter. The outer green pericarp is about 3 mm thick and fibrous in construction, dehiscing along the ventral suture at maturity, exposing the nut. The nut is a true seed, with seed coat, hilum and micropyle. The latter may be open or closed. The seed coat, or testa, is 2 to 5 mm thick and is very hard to brittle. It encloses the kernel or embryo, which consists of two large cotyledons and a minute plumule and radicle.

Work by Jones and Shaw (1943) on nut development showed that the shell and husk develop early. The embryo sac at this time contains a gelatinous endosperm. The shell is hard within 111 days. The embryo or kernel makes little growth before 90 days and little oil is accumulated. By 140 days the embryo replaces the endosperm. Dry weight of kernel, represented mainly by oil, increases continually from 90 days to 215 days, or when close to maturity. Seventy percent of the oil, however, is formed between 90 and 185 days (44% of the growth period).

Jones and Shaw (1943) suggest carbohydrate is the most likely source of oil. It would initially come from starch reserves in flowering branches. However, they state that in such branches starch reserves are largely depleted by the time oil formation begins. This begs the need to investigate the source of carbohydrate at this stage and the influence this has on nut quality which is correlated with oil content.

The composition of nuts has been reviewed by many authors and it varies during nut formation and between species and varieties at maturity. In broad terms the dried nuts contain approximately 73% oil, 6% sugars, 9% protein, and no starch. They are also a source of calcium, phosphorus and the vitamins thiamine, riboflavin and niacin.

Climatic requirements

M. integrifolia is found in coastal and tropical rainforests on the eastern slopes of the Great Dividing Range in Queensland from Mt Bauple, on the lower Mary River near Maryborough, south to lower Beechmont or the Numinbah Valley, a distance of about 440 km; latitudinal range, approximately 25° to 28°S.

M. tetraphylla is found in similar country but more south, extending from the Coomera River near Mt Wongawallen, and the Nerang River near Advancetown to the Richmond River near Casino and Lismore in north-east New South Wales, a distance of 120 km; latitudinal range, approximately 28° to 29°S.

At a point where these species meet, there are types which appear to be natural hybrids.

The distribution of the species in their natural state gives some idea of climatic requirements but work in other crops has shown this to have limitations. This is further indicated by apparently better yields from *M. integrifolia* selections in Hawaii than Australia.

A broad evaluation between the two areas shows that the Hawaiian climate is more equitable both in temperature and rainfall. For maximum production of *M. integrifolia* in Hawaii, a climate with an absolute maximum and minimum of 32°C and 9°C respectively is required. This, however, gives poor performance of *M. tetraphylla* and hence the lack of interest in that species in Hawaii. Rainfall in

Hawaiian production areas mostly exceeds 2000 mm, which is higher than in most Australian localities, and distribution is fairly even, in contrast with the Australian 'winter-spring' drought. The comparison may be questioned, as it is based on Hawaiian varietal selections. Although one may expect local selections adapted to Australan conditions to perform better, no selections to date have shown much improvement over Hawaiian selections.

High temperatures for extended periods obviously affect the growth on the northern sides of trees in summer months, in coastal areas north of Bundaberg. Variety 508 is one that shows high susceptibility.

Macadamias can withstand light frosts and in California *M. tetraphylla* has withstood –3°C for long periods. In practice, frost-free areas should be chosen, as frost readily kills trees. It has been shown that there is genetic variability in temperature susceptibility at both ends of the scale in both species.

Humidity is not easily related to yield or quality, but has been related to disease attack on flowers. In Hawaii *Botrytis cinerea* has reduced yields through loss of flowers, when leaves and flowers have been wet for extended periods at flowering. Continual heavy rain at flowering has the potential to remove pollen and could be significant where cross-pollination is important and bee activity is reduced.

Nakata (1976) shows that night temperature influences floral development. Initiation occurred at 12, 15, 18 and 21°C, with the greatest production of the most racemes occurring at 18°C. Recently Stephenson (unpublished data) has shown that flower initiation occurs in May, in southern Queensland. The night temperature at this time apparently could limit the performance in two ways. Firstly, in high temperature localities poor flowering may result and secondly, as may be the case in many Australian production areas, over-flowering may tax reserves important to nut retention.

Shigeura *et al.* (1975) have shown that elevations above 500 m and 450 m affected yield and quality respectively. This is thought to be due to lower prevailing temperatures during the nut development period.

Irrigation has been shown by several workers to boost yields in drought years, and there is little doubt that in lower rainfall areas or areas of poor rainfall distribution, irrigation is essential.

Radspinner (1971) showed that high rainfall and temperatures in the two months before flowering reduced kernel quality, nut size and kernel recovery, while increased rainfall during the period of oil formation increased kernel weight.

The effect of light on tree performance has been demonstrated, and thinning of crowded trees dramatically increases the yield of the remaining trees. As trees grow the nut-carrying area follows the canopy out, decreasing the bearing capacity at the tree centre.

No information is available on photoperiodism.

Soil requirements

Experience has shown that macadamias will perform on a wide range of soil types from open sands and lava rock soils to heavy clay soils. They are however affected by waterlogging. Restricted drainage impedes root development, hence tree growth and performance. In practice, soils of high fertility, and well drained to at least 0.5 m, are preferred. Greater response has been recorded with organic matter additions on sandy soils, hence soil organic levels are important.

Genetic diversity

Genetic diversity is based on the ready compatability of the two edible species. Within the species, characteristics range widely. There is an overlapping between species of some characteristics.

M. tetraphylla is recognized as peforming under colder conditions than *M. integrifolia.* The latter is used commercially, well into the area of natural distribution of the former species. Work has shown that selected *M. tetraphylla* trees grow well in the heat at Rockhampton, while the *M. integrifolia* selection 508 suffers from sun scald. This diversity and overlapping of characteristics offers wide opportunity for plant improvement.

Nut quality standards and processing characteristics have been set on *M. integrifolia.* Because of this direction there is presently a reluctance to incorporate *M. tetraphylla* because of the probable variation from the standard.

Within these species and particularly *M. ternifolia* there is a bitter element, a cyanogenic glycoside, proteacin. The element appears at different levels even in acceptable kernels and rises to 0.8 mg per gram in unpalatable nuts. The level is also under some environmental control and nuts must be checked over several seasons when selecting varieties. Only 30% of kernels on a given tree are affected. The problem does not exist in any of the named Hawaiian varieties.

Selection of macadamia varieties is based on several nut quality characteristics on which there is general agreement. Briefly these characteristics are as follows:

- Total kernel recovery: the total weight of kernel expressed as a percentage of the total nut-in-shell weight in a laboratory sample.
- Commercial kernel recovery: the total weight of sound and acceptable-size pieces of kernel following commercial cracking, expressed as a percentage of the total nut-in-shell weight of a sample.
  (These two recoveries should be expressed at a specific moisture content.)
- Percentage of no. 1 grade kernels: the percentage by weight of kernels which float in tap water. Such kernels have 72% or more oil. This character is a fair indicator of cooking quality and storage life.
- Size: Kernels within the range of 2 to 3 gm (at 1% moisture) are acceptable.
- Shape: a uniform smooth kernel is desirable, as abrasion is less likely to damage the appearance and cause a loss of A-grade kernel percentage.
- Colour: a creamy white uniform colour is desired for kernels. Any discoloration is not acceptable.

Hamilton *et al.* (1968) report two areas of work which show that these characteristics do vary with locality for a given variety. However, the relativity between varieties remains constant even when comparing varietal performance between Hawaii and Australia.

Hamilton and Fukunaga (1973) claim that it is feasible to further increase, by plant improvement, production per tree by 25%, kernel recovery by 45%, and percent no. 1 kernels by about 90% up to 95%. This potential if realized could lead to an increase in production of no. 1 kernels by up to 70% over the mean for the seven named and approved Hawaiian varieties.

Selection characters which cause rejection are stick tight nuts and open micropyle. Upright growth in trees is a character sought after because of its reputed higher potential yields per unit area.

The characteristics of the varieties which are grown commercially in Australia are set out in the table opposite. The figures quoted are drawn from various sources with variable standards and should be taken as a guide only.

Characteristics of macadamia cultivars grown in Australia.

Variety	Hawaiian name	% kernel recovery	Nuts per kg	% no. 1 kernels	Tree shape	Notes
508	Kakea	35–39	143	90	Round to conical	Not suited to hot areas.
246	Keauhou	37–40	139	85	Broad and spreading	Losing favour to newer varieties.
333	Ikaika	31–35	154	89	Round to conical	Nut quality suspect. Wind resistant type.
660	Keaau	42–46	176	97	Upright conical	Preferred type for close plantings.
344	Kau	35–39	132	98	Upright conical	Preferred type for close plantings.
800	Mackai	40	125	97	Round	Suggested for trial in warmer areas.
741	Mauka	43	154	98	Upright conical	Suggested for trial in cooler areas.
Own choice	—	35–39	130	95	Round and dense	Has stick-tight nuts. Not preferred.
H2	—	32–36	143	85	Upright vase shape	Has stick-tight nuts. Not preferred.

Production season and time of cropping

Flowering in macadamia depends on the cultivar. Some cultivars have an extended flowering period which may have peaks resulting in nuts of varying maturity times. In south-east Queensland racemes will appear on trees for up to 4 months with peak anthesis ranging from July to September. Nut maturity depends on locality, variety and year. A guide from work by Mason (1979) showed a general trend of maturity by district: Bundaberg — mid-February, Rockhampton — early March, and Maleny — mid-April. Variety 246 usually matures in advance of 508 by two weeks. A two-week difference was also recorded between two years of recordings for optimum maturity time. Very few nuts fall before maturity, and fall may extend until June. There is opportunity for use of tree shakers or nut drop-inducing chemicals to compact the harvest period.

Yield potential

Yield potential is best stated on a per hectare basis although most reports are on a tree basis. As yields depend greatly on spacing and inter tree competition, such tree yields are misleading. Yields in Hawaii are recognized to exceed commonly recorded Australian yields by a margin of 30 to 50%. Yields in Hawaii are stated to be between 7000 to 8000 kg per hectare nut-in-shell (10% moisture). Australian top yields are approximately 5500 kg per hectare nut-in-shell (10% moisture). These figures show an opportunity for further development in Australia. It is considered that environmental limitations are the major factors controlling yield with the present selected varieties lacking adaptation to the prevailing conditions.

Market acceptability

The majority of nuts are sold as processed kernels in the confectionery trade. The macadamia is sold and maintains a status as a high-quality nut. Even at the high prices demanded, twice that of cashew, markets are still expanding.

Markets exist for various forms of combined confectionery, such as chocolate and candy, and also in the ice cream and prepared dessert trade.

As a table nut-in-shell there is only limited demand as most varieties are too hard to crack manually. Tetraphylla types are available which can be easily cracked but there appears to be only a limited market to a select group of enthusiasts.

There is a market for macadamia oil at a price. The shell has been tested as a filler in the plastics industry. The husk has a place in the nursery and landscaping trade.

Export potential

Export markets have already been established and their development is limited by supply. The world nut market is said by Axer (1980) to be 680 000 tonnes. Compared with almonds 35%, cashews 14%, and pecans 6.4%, macadamias are grouped with the also-rans below 1.0% in the world market race. The eating quality of the macadamia excels that of most, if not all its competitors. There is obviously immense opportunity for expansion with good marketing in affluent societies of the world.

Cultural details

Propagation

Techniques for propagation are well developed. Both grafting and budding of seedling rootstocks are used. The wood of macadamia is hard, requiring the propagator to have experience to be successful.

Grafting is recommended in spring or autumn. The scionwood is cinctured some six to eight weeks beforehand. Preferred wood is healthy mature material of the previous flush. The recommended graft is the simple whip, using material 1 to 1.3 cm thick. Often a small wood plane is used to gain even surfaces on the hard wood of the stock and scion.

The side graft is used in Hawaii with success. Tip, wedge or cleft grafting is used under glasshouse conditions for working small seedlings up to 30 cm high. Seed grafting has been used where the plumule was wrenched from the germinating seed and the wedge type scion forced into the soft remaining tissue between the coytledons. The method has lost favour, as trees are thought to develop poor root systems and are subject to being blown out in windy locations.

Budding is gaining favour and is especially useful where new varieties are being multiplied. The punch bud, where the bud is removed and similarly inserted on the stock with a very sharp rounded to oblong punch, is the most popular technique. The bark must lift readily on stock and budwood, and speed and care are required to prevent dehydration. The patch and chip bud techniques have also been successful.

Management

PRE-PLANT.

Plantations are now designed for mechanical harvesting so land preparation and layout design should take this into consideration. Land should be levelled to the state where equipment can be used efficiently. Removal of surface stone is also important. Long continuous rows within reason are desirable.

Stumps and large roots should be fully removed as these harbour armillaria fungus which will later attack macadamia. Minimum cultivation is required and only machinery hardpans require rupturing by ripping. The desired minimum soil pH is 5.5. Soil analysis is recommended to guide adjustment for this and other elements such as phosphorus, calcium, magnesium, copper and zinc before planting. These elements require mechanical incorporation to be effective so this is best done before planting.

PLANTING.

Autumn or spring plantings are suggested to avoid extremes of temperature. Macadamia has a high water requirement and should not be planted until a permanent irrigation system is in operation.

Hardened plants establish best. Trees should be positioned with the bud or scion towards the prevailing wind, and with a 10% lean in that direction in windy positions. Staking has lost favour as it has been found that self-supporting trees develop a more rigid trunk better able to support itself. Tree guards for wind protection increase growth rate but are not recommended because of the high capital and labour cost.

Wind-breaks are desirable as the wind burns new vegetative flushes and restricts growth. Permanent tree wind-breaks such as eucalypts are suggested on ridge tops and edges of plantations. The alternative is to plant taller growing (344) or wind resistant macadamia varieties (333) in these positions. In the establishment phase, five to six years, interrow plantings at every third row of a narrow but medium height break, such as barner grass, is recommended. Intensive sheltering has increased the incidence of pests in some locations. In cooler areas free flow of cold air should not be impeded as frost incidence and damage can be increased.

The spacing adopted depends greatly on return to capital in the first 10 to 15 years. It also depends on access of spraying and harvesting equipment. Layout also depends on equipment access and irrigation system design. The direction of the hedgerow should be dependent on suitability for harvesting and irrigation design, rather than orientation to the sun. Interrow distances are 10 to 11 m with adjustment for potential varietal size and induced vigour from local environment. In-row distance is 4 to 5 m, also dependent on the above factors.

TREE MANAGEMENT.

Tree management should aim to gain maximum growth before flowering and cropping commences, in order to achieve early returns for investment. Yield is related to tree size and exposed canopy.

A study of young tree training by Hamilton and Fukunaga (1959) describes a detailed method of both obtaining good limb placement, and preventing acute crotch angles. With the advent of mechanical harvesting since this study, thought must also be given to machinery movement.

In practice it is found that the detailed method is not adopted by commercial growers because of its complexity. In addition Trochoulias (personal communication) has found that any pruning of trees significantly reduces growth rate and increases time of flowering. Presently growers within the first 2 years after planting aim only to remove major branches which develop poor crotch angles. These would jeopardize the future structural stability of the tree. Lower limbs are not removed unless there is a problem with mechanical harvesting.

While trees continue to expand, yield rises until crowding and reduced light interception off-set potential yield. At this stage there are a number of choices of tree management in the row: either remove trees at this stage, at some later stage, or leave a full hedgerow to develop. The last of these is most favoured because it avoids interruption in early returns, the mental anguish of removing bearing trees, the cost of tree pruning and tree removal, and the potential for increased wind damage in a newly opened orchard.

Work on thinning trees by Hamilton *et al.* (1978) showed that gross tree yield improvements of 100% could be attained by thinning from 8 m x 8 m to 11 m x 11 m. However, yields per unit area did not increase significantly, which is possibly of greater interest in tree management decisions.

Management of the interrow space requires first the decision either to leave sufficient space for mature trees, or to adopt tree pruning. There is no evidence on which a sound decision can be made in this regard. Liang and Myers (1976) reported that tree pruning caused a reduction in yield, somewhat equivalent to the amount of wood removed. Mutual shading of trees in this plane would also give similar reductions in yields. If interrow competition can be off-set by sufficient spacing, it appears the problem can be limited. It is believed 10 to 11 m is ideal. If closer spacings are adopted, pruning may be the lesser of the two evils when canopies close.

Irrigation

Awada *et al.* (1967) and Trochoulias (personal communication) have shown that irrigation gives yield responses. In each case, however, the response has been variable from year to year and unpredictable.

Water stress is believed to affect nut quality. It is difficult to isolate this factor as consideration must be given to premature nut fall. This would camouflage the effect of stress during nut development as premature nuts will be of low quality. It is thought that stress at specific periods in the crop cycle must be more important than at others, to give the erratic response received from irrigation.

Present recommendations suggest high rates of watering during the establishment phase to achieve maximum plant growth. From planting, trees for the first 2 months receive 20 to 30 L of water per tree, twice per week. In the first year, weekly watering is required when rain does not occur. Rates during this period will rise to 130 to 150 L per tree per week in hotter months. During winter months this rate falls to about one quarter. Maximum rates per tree per week for about 10 years increase by the above rate for each year of age, that is, 5 years = 750 L. Application rates for trees older than 10 years will depend on spacings used and should be adjusted proportionately for tree size. These figures are only a guide and should be adjusted to meet local conditions.

Critical periods in the crop cycle are believed to be from the time of nut set, through nut filling, and through the vegetative growth period in midsummer.

Fertilization

A wide range of research work has been reported on major and minor elements mainly emanating from Hawaii. This work has lead to the identification of optimal leaf levels. N — 1.50%, P — 0.075%, K — 0.45%, Mg — 0.10%, Ca — 0.6–0.9%, S — 0.24%, Zn — 15 ppm, Cu — 4.5 ppm, Mn — 100 ppm, B — 75 ppm.

Responses have been recorded visually or in growth or yield for N, P, K, Zn, B, S, Mg, Fe and Cu. This indicates the macadamia is very responsive to nutrition. Experience in Australia confirms this situation, and also that better yielding trees relate well to the leaf levels stated above.

Fertilizer rates and schedules from Hawaii are coloured by its high rainfall and low fertility lava soils. They cannot be directly adopted in Australia. The study of the use of leaf levels is also in its infancy in Australia. Although they are useful as a guide for major problems, their use for close management of fertilizer rates is not established.

In the commercial Australian scene, leaf and soil analysis services are available which supply recommendations based on the history of the plot and the related analysis. A basic fertilizer programme in vogue is 120 g urea, 50 g superphosphate and 80 g muriate of potash per tree per year of age up to 10 years of age. After this stage, the spacing controls tree size, and the tree rate is increased proportionately with increases in tree size. Suggested application times are, in years 1 to 5, five equal split applications in February, April, July, October and December of nitrogen and potassium. In older trees the application is split: nitrogen, $^1/_2$ February, $^1/_4$ April, $^1/_4$ October; potassium, $^1/_3$ February, $^1/_3$ April, and $^1/_3$ October, Phosphorus in each case is applied in one application towards the end of the wet season. Such a recommendation is only a guide as it is based on very limited trial work.

Pests and diseases

Predominantly foliage pests:

- Macadamia leafminer *(Acrocercops chionosema)*
- Macadamia leafminder *(Acrocercops chionosema)*
- Macadamia twig-girdler *(Neodrepta luteotactella)*

- Redshouldered leaf beetle *(Monolepta australis)*
- Orange fruitborer *(Isotenes miserana)*
- Macadamia cup moth *(Comana fasciata)*

Predominantly flower pests:
- Macadamia flower caterpillar *(Homoeosoma vagella)*
- Hairyline blue butterfly *(Erysichton lineata lineata)*
- Macadamia lace bug *(Ulonemia* sp.*)*
- Black citrus aphid *(Toxoptera aurantii)*

Predominantly nut pests:
- Fruitspotting bug *(Amblypelta nitida)*
- Bananaspotting bug *(Amblypelta lutescens lutescens)*
- Macadamia nutborer *(Cryptophlebia ombrodelta)*
- Yellow peach moth *(Dichocrocis punctiferalis)*

Diseases:
- Macadamia trunk canker *(Phytophthora cinnamomi)*
- Minor diseases include armillaria root rot, raceme blight *(Botrytis cinerea),* nut anthracnose *(Colletotrichum gloeosporioides* var. *minor)* and a cercospora husk spot.

Good management and control measures are available for all these problems. They are described in the Queensland Department of Primary Industries Horticulture Branch Advisory Leaflet H38, 'Pests and diseases of macadamia nut', February 1980.

Weed control

General weed control is maintained with paraquat (Gramoxone R). Care must be taken to avoid contact with stems as damage and entrance of phytophthora trunk canker can result. Glyphosate (Roundup (R)) is used, although not presently registered for use, to remove perennial grasses and resistant species not controlled by paraquat. Trials have been carried out on diuron, and trees on medium clay soils have shown good tolerance, but it is not used in the industry.

Harvesting methods

The nut, because of its hard shell, is suited to mechanical harvesting. Hand harvesting costs are high. Orchards should be levelled and designed to accommodate commercial equipment which is available. A range of American 'sweepers' and large harvesters, used in other nut crops, are applicable to the macadamia and are in use. The cost of such equipment makes them only cost efficient in orchards over 25 ha. Newly developed Australian equipment is now on sale for smaller enterprises.

Nuts can be either harvested following natural fall, treatment with ethrel, and/or shaking to induce fall. Commercially, natural fall is mostly utilized. Possible cost benefits, management, and side effects of the other two methods require further study and development if they are to have wider acceptance.

Nuts fall with the husk attached. When harvested they are transported to a central orchard facility where the husk is removed by machinery before drying and storage. Drying is done in racks or forced air silos to bring the moisture to at least 10% before sale. Further drying to 1.5% moisture using heated air (52°C down to 4.5% moisture and then 77°C down to 1.5%) is required for long term storage, cracking and processing. Excessive heating will damage nut quality. Details of equipment and procedures are readily available to industry.

Marketing, storage, and transport characteristics

Nuts are usually marketed to processors as nut-in-shell. In this condition they can be bulk handled and mechanically transferred. Processors store such nuts at around 1.5% moisture in silos without spoilage for periods in excess of 1 year.

Processors sell kernels after cracking and grading to wholesalers as either a processed or raw product. Processing is either by deep frying in refined deodorized coconut oil or by dry roasting. Salt may be added. In each case the product is sealed in vacuum packs of metal, foil or other flexible materials. The storage life of such products is considered to exceed 1 year provided the basic product was sound, had reached the prescribed 72% oil content, and the processing and packing techniques were satisfactory.

Processing characteristics

The general characteristics required for processing are given under varietal selection in the section Genetic Diversity. A detailed assessment of recent industry requirements and methods of handling processed nuts is given by Nissen and Williams (1981).

References

Awada, M., Warner, R.M., and Watanabe, Y. (1967), Effects of soil moisture stress on growth and yield of macadamia. Hawaii Agric. Expt. Station, University of Hawaii Technical Bulletin 69.

Axer, J. (1980), The world of nuts. *American Fruit Grower Journal*, October.

Hamilton, R.A., Cameron, J.W., and Mouat, H.M., (1968), Nut characteristics of macadamia varieties remain relatively stable at different locations. 8th Annual Proceedings Hawaii Macadamia Producers Association.

Hamilton, R.A., and Fukunaga, E.T. (1959), Growing macadamia nuts in Hawaii. Hawaii Agric. Expt. Station, University of Hawaii Bulletin 121.

Hamilton, R.A. and Fukunaga, E.T. (1973), Development and evaluation of macadamia nut varieties. California Macadamia Society Yearbook.

Hamilton, R.A., Fukunaga, E.T., and Ito, P.J. (1978), The effect of thinning and increased spacing on yields of Kakea and Keauhou at Kona Branch Station. 18th Annual Proceedings Hawaii Macadamia Producers Association.

Jones, W.W. and Shaw, L. (1943), The process of oil formation and accumulation in the macadamia. *Plant Physiology* **18,** 1.

Liang, T. and Myers, A.L. (1976), Macadamia harvesting research beyond tree shaking. 16th Annual Proceedings Hawaii Macadamia Producers Association, p. 46.

Malcolm, H. and Trochoulias, T. (1979), Proteoid roots help macadamia nut trees. *Agricultural Gazette of New South Wales* vol. 90, no. 1, Feb.

Mason, R.L. (1979), Macadamia nut maturity studies. Australian Macadamia Society News Bull. vol. 5, no. 6, April.

Nakata, S. (1976), Progress report on flowering, nut setting and harvesting, with special reference to the effects of night temperature and growth regulators. 16th Annual Proceedings Hawaii Macadamia Producers Association.

Nissen, R.J. and Williams, R.R. (1981), Assessment of macadamia selections based on nut quality. Australian Macadamia Society News Bull. vol. 8, no. 4, Oct.

Radspinner, A.L. (1971), A study of variability of *Macadamia integrifolia*.Ph.D. thesis, Univ. Hawaii.

Shigeura, G.T., Bullock, R.M., Silva, J.A., Harada, W. and Sumida, D.A. (1975), Some factors affecting growth, production and nut quality in macadamia nuts. 15th Annual Proceedings Hawaii Macadamia Producers Association.

Smith, L.S. (1956), New species of and notes on Queensland plants. *Proceedings of the Royal Society of Queensland* **67,** 39.

Urata, V. (1954), Pollination requirements of macadamia. HAES Technical Bulletin 22.

Further reading

Johnson, L.A.S., (1954) *Macadamia ternifolia* F. Muell and a related new species. *Proceedings of the Linnean Society of New South Wales* **79,** 15.

Jones, W.W. (1939), A study of development changes in composition of the Macadamia. *Plant Physiology* **14,** 755.

40 Carambola

41 Coconut

42 Passionfruit

43 Sweet Granadilla Flower

44 Granadilla

45 Sweet Granadilla

46 Macadamia

47 Macadamia

48 Coffee

49 Chinese Jujube

50 Casimiroa

51 Pummelo

Chapter 23

RHAMNACEAE

CHINESE JUJUBE
Ziziphus jujuba Mill.

by D. McE. ALEXANDER

INDIAN JUJUBE
Ziziphus mauritiana Lam.

Botany

Included in this family, which has 46 genera and over 500 species of woody trees, shrubs and climbers, are two species of *Ziziphus* which are grown commercially. These originated in northern China *(Z. jujuba* Mill.*)* and in Burma or northern India *(Z. mauritiana* Lam.*)* (De Candolle 1964). According to Popenoe (1920) *Z. vulgaris* Lam. and *Z. sativa* Gaertn, are invalid names for *Z. jujuba* Mill.

Both *Z. jujuba* and *Z. mauritiana* are grown over a very wide area. For *Z. jujuba* this extends from northern China to Syria and countries bordering the western Mediterranean where it grows wild. *Z. mauritiana* is grown in warmer parts of India including the Punjab, Haryana, Rajasthan and Uttar Pradesh. The Chinese date is known locally as tsao while the Indian jujube is known locally as ber. *Z. jujuba* has been cultivated in northern China for more than 4000 years. It is a large shrub or small deciduous tree up to 12 m high which is very cold tolerant but has a very low chilling requirement. It is generally spiny but some spine-free cultivars are known. The glossy leaves are alternate, unifoliate, oval or oblong and 3 to 8 cm long. Flowers are small $1/2$ cm to 1 cm), white or greenish petalled and open in late spring (about 3 months after peaches). The fruit is a drupe up to 5 cm long, oval, and when ripe has a thin, tough mottled brown or mahogany-coloured skin enclosing a sweet, crisp whitish flesh resembling that of an apple. It ripens in autumn.

If kept on the tree the skin turns dark brown and the fruit begins to shrivel. Dried, these are the Chinese dates of commerce. The single seed is woody and resembles that of a date but is pointed at the distal end. Cultivated forms are sometimes thornless and have larger and sometimes sweeter fruits. *Z. mauritiana* differs from *Z. jujuba* in having a cream-coloured, downy underside to its leaves. It is also less frost hardy, and thrives in a wet or dry tropical climate.

Climatic requirements

Both species can withstand hot, desert conditions. They are cold tolerant (to –22°C) and, because they are late flowering, can withstand spring frosts. Long, hot, dry summers with plenty of sunshine are required to properly ripen the fruit. High humidity limits satisfactory fruiting of ber in many parts of India.

In India, ber trees have a dormant period and shed their leaves in May or June, after the fruit harvest. Growth recommences with the Wet in July and most growth occurs before November when it is checked by cold.

Soil requirements

Z. jujuba prefers a sandy loam which is neutral or slightly alkaline. It can be grown on heavy textured, poorly drained soils and on soils which are slightly acid. Trees can withstand high salinity, drought and waterlogging.

Genetic diversity

A great number of varieties of both species is cultivated. In California the *Z. jujuba* varieties Lang and Li are grown. The Lang has large, rather pear shaped fruit while the Li fruit is round with a very small seed. In India, Singh *et al.* (1973) have recommended eight varieties of ber for commercial production, namely Umran, Sanaur No. 1, Sanaur No. 2, Kaithli, Dandan, Illaichi, Muria Mehrara and Gola.

Production season and time of cropping

Trees of both species crop at an early age, but it is recommended that for the first few years fruit be removed in order to develop a larger tree. *Z. jujuba* fruit matures in autumn while that of *Z. mauritiana* matures in March and April (spring) in India.

Yield potential

Jujubes produce from 100 to 250 kg fruit per tree at tree spacings of about 9 m between trees (120 trees per ha).

Market acceptability

Jujubes can be used fresh, dried or candied. Fruit composition resembles that of figs and dates. Figs contain more fat, jujubes more protein and dates more sugar than the other two fruits.

Export potential

There is possibly some export potential for high quality fruit and for processed fruit, but it must be remembered that dates would be the major competitor.

Cultural details

Propagation

Z. jujuba is propagated from seeds, cuttings, buds and grafts. Seedling or root cuttings are used as rootstocks.

Grafting with scion wood about pencil thickness, using either a whip or whip and tongue, has been the most satisfactory method. Seeds of small fruited, wild types often give better germination than those of selected commercial cultivars. *Z. mauritiana* has a long tap-root and is therefore difficult to transplant. Seedlings are pruned to one shoot and budded by either the T-bud or ring-bud method. In India best results are obtained in July, August and September. In Uttar Pradesh air layering has been practised. Two-year-old shoots from mature trees are treated with 10 000 ppm indole butyric acid and napthalene acetic acid in lanolin paste, prior to wrapping with sphagnum moss, to give a profusion of roots in 2 weeks. Cutting propagation has also been successful.

Management

Early pruning is recommended to shape the trees and to produce a spreading rather than an upright tree. In India young ber trees are supported by bamboos, lower branches to 1 m are removed, and four to five branches are selected as scaffolding. In subsequent years 25 to 50% of the shoots are removed annually when the trees are dormant. Approximately 120 trees are planted per hectare.

Pests and diseases

Jujubes are generally free of pests and diseases but in India fruit fly, lac insects, defoliating beetles, termites and *Oidium* sp. have been reported. Birds can also be a problem as fruits ripen.

Harvesting methods

Hand harvesting is practised but for drying, fruits can be collected after they fall to the ground.

Marketing characteristics

Fruit is marketed fresh, dried or candied. For the latter the skin is punctured and the fruit is either soaked or boiled in a sugar solution and then air dried.

Storage characteristics

Dried or candied fruit can be stored for several years. Fresh fruit is readily transported and dries out if kept in a dry atmosphere.

Transport characteristics

Because of its tough skin jujubes are readily transported without damage. Dried or candied fruit is readily transported.

Processing characteristics

Fruit for drying is scored with a multiple bladed knife or dipped in boiling water to break the skin before exposure to the sun. For candying the skin is punctured with multiple wooden or metal spikes to aid the movement of the sugar.

References

Sing, K.K., Chadha, K.L. and Gupta, M.R. (1973), 'Ber'. Punjab Ag. Univ., Ladhiana.

De Candolle, A. (1964), 'Origin of cultivated plants'. Hafner, New York.

Popenoe, W. (1920), 'Manual of tropical and sub-tropical fruits', a facsimile of the 1920 edition, published by Hafner Press, New York, 1974.

Further reading

Lee. B.W. (1975). The jujube or 'Chinese date'. Leaflet 2729. Univ. Calif., Berkeley.

Chapter 24

RUBIACEAE

COFFEE

by B.W. CULL

Coffea arabica L. and **Coffea canephora** Pierre ex Froehn.

Coffee is an established world crop grown mainly in presently low–cost-labour countries. This review aims to identify the factors which might be pertinent to the adoption of the crop in Australia. Emphasis is on those factors most significantly pertaining to plant management and mechanisation of the crop with high labour costs.

Botany

There are probably about 60 species of *Coffea*. Within commercial coffee the major species are *C. arabica* L., *C. canephora* Pierre ex Froehn. and to a very limited scale *C. liberica* (Bull ex Hiern.). *C. arabica* and *C. canephora* will be referred to as Arabica and Robusta respectively in the text. This covers the two general types identified by the trade.

The botany discussed will be that of Arabica. Robusta is generally similar and only the major variations will be covered.

Of immediate botanical importance is the stem and branch structure of the plant and the growth pattern, and then the flowering structure and its initiation and development.

Coffee exhibits dimorphic branching in that it has an upright vegetative (orthotrophic) stem from which arise lateral or primary (plagiotrophic) branches. The orthotrophic stem rarely carries flowers while the plagiotrophic branches are the major fruiting source. Both forms arise as buds one above the other in the leaf axils, the latter being above.

The primaries arise on the opposite side of each node in succession from the base upwards around the stem. They are held in variety Arabica at 90° to the main stem. However, Bourbon variety is more upright at 55°. When the primaries die or are cut back to the upright stem there is no other bud to replace it. This necessitates the replacement of the upright. This can be done by cutting back to above a node. The lower buds of that node then move away to form two new leaders which are vegetative.

Thus vegetative propagation (cuttings and grafting) must be performed using only the orthotrophic stems to get a normal plant. Cuttings of primaries develop a weak rambling plant which never revert to the upright vegetative form.

In the axils of leaves on primary or fruiting branches there is a series of 6 buds borne one above the other. The top bud distal from the petiole is the largest and first to respond. Under conditions conducive to flowering the first 3 or 4 buds usually develop into inflorescences. They can, however, all form into vegetative shoots by damage or pruning of the primary branch beyond them giving rise to secondary and ultimately tertiary fruiting branches. Under the influence of insects, or cold damage, or boron deficiency a prolific fan of branches will form which is highly undesirable.

The production of good quality coffee and the maintenance of a tree structure suitable for mechanization depends on the control of the branching structure of the plant. Evidently environmental and cultural practices can disrupt the balance.

Flowers are carried on the mature wood of the previous growth cycle but can occur on older wood. The inflorescences in the main are produced on 1 to 6 buds as described above. Each inflorescence consists of 4 flowers. Each possible bud and flower need not develop and 2 to 20 fruit per axil are common. Usually 40% of flowers mature fruit, the major loss of fruit occurring in the first 13 weeks of the total 30 to 35 weeks taken to mature the berries (Robusta takes 40 to 44 weeks). Size is determined in the first 17 weeks. The fruit are a very dominant sink and resist fall strongly, but they can easily sap reserves which may result in damage to the plant or an alternate bearing habit.

Coffee is considered a short-day plant, setting more flowers under 8 to 10-hour days than under 12 to 13-hour days. Better flowering has been recorded for relatively lower temperatures; 23 to 17°C day performed better than 26 to 20°C or 30 to 24°C. High temperatures are related to the development of sterile 'star flowers' which are green. It is claimed also that 2 to 3 months of dry weather is needed for initiation. Water stress does actively keep flower buds dormant in spring. Flowers open 9 days after 8 mm of rain. The rate of opening depends on the temperature.

Early flowering in spring can retard vegetative growth. Yield, quality and survival of the plant is largely dependent on the leaf area to crop ratio. The inability of the plant to make sufficient vegetative growth impedes carbohydrate accumulation and also actual growth on which the following season's crop is carried. It is possibly an advantage to delay flowering as late as possible, firstly because of the need to gain vegetative growth but also to get the most concentrated period of flowering which will concentrate the crop and the harvest period.

The flowers are carried on short peduncles and the ovary is composed of two united uni-locular carpels, each containing a single ovule. The floral structure allows cross-pollination but self-pollination is more common in Arabica with recordings of only 7 to 9% cross-pollination occurring. Robusta is self-sterile hence cross-pollination is required. In vegetatively propagated material two lines of Robusta must be integrated to get pollination and fruit set.

The fruit is a drupe and its structure causes problems in processing. The pericarp is differentiated into 3 definite layers, a tough outer exocarp enclosing a pulpy mesocarp and a definite hard endocarp (parchment) surrounding the two seeds. The seeds are deeply grooved and covered with a thin testa (silverskin) which penetrates the groove. The removal of the outer coatings, particularly the silverskin, requires a certain expertise and is costly.

Different sources quote that 4 to 6 kg of berries yield 1 kg of dried beans.

Climatic requirements

The origin of Arabica coffee is Ethiopia in a forest understory at approximately 1500 to 2000 m above sea level. Robusta is more equatorial being found between 10°N and S and from the west coast of Africa to Uganda and between sea level and 1600 m.

Arabica is grown south to 24°S in Brazil. Frost is a major limiting factor in that it kills plants. The optimum temperature range is quoted as 16 to 23°C. Photosynthesis tends to fall above 23°C until it reaches nil at 34°C. The plant is capable of withstanding higher temperatures but high temperatures may cause apical dominance to break down and upright suckers to form.

Prolonged periods below 12°C may affect growth and development. Growing areas in Brazil experience 2 months (June and July) with a mean monthly temperature of 15°C. In this period the plants are in a state of desired dormancy so these low temperatures may not significantly affect yield. Cold temperatures, but no indication to what degree, are reported to induce excessive secondary and tertiary branching. This is possibly cultivar related and can only be tested by local trial.

Coffee is grown in areas with 900 to 2000 mm of rainfall. Under Australian conditions where maximum control is to be attempted with flower development and cropping, the lower rainfall areas with a high probability of no rain in June to

September are probably ideal. This would also assist harvesting operations. Provision of irrigation would be essential.

Wind protection is desirable but this is no problem with the available wind-breaks.

Several localities in Australia would meet the frost-free, low rainfall requirements. Arabica coffee is being grown well in South Africa in areas which equate closely to most of the frost-free areas on subcoastal tableland and range country along the Queensland coast. The effect of high and low average mean temperatures would require investigation.

Soil requirements

The literature indicates that deep well-drained soils are ideal for coffee. Where control is required over growth and flowering, soils which have a lower water holding capacity and low ability to retain nitrogen are possibly more easily managed in this respect.

Genetic diversity

The genetic diversity in coffee is wide and apparently has been only partially exploited.

Coffea arabica has two botanical varieties Arabica and Bourbon. The former was the material which reached the Americas by way of the Dutch and subsequently the French. At one stage in this devious route the population was reduced to a single plant. The genetic variability on which the American industry was originally based was therefore limited.

Bourbon was taken to Reunion (previously Bourbon) by the French from Ethiopia and then on to Africa. This Arabica variety has greater diversity. Today many mutants of both varieties and crosses between them are recognized as cultivars. Within these cultivars is resistance to many of the important diseases includng *Hemileia vastatrix* leaf rust. Rootstocks are recognized by resistance to nematodes. High yields, dwarfing, tree shape, branch angle, and concentration of flowering time are all recorded as variants.

Interspecific hybrids are possible, but although progeny occur none are used in commerce.

It is of interest that Arabica is tetraploid and Robusta is diploid. Hybrids form triploids which are sterile. Natural mutants of Arabica can form hexaploids and octoploids which are apparently poor producers.

Production season and time of cropping

Under Australian conditions the general plant growth and flowering cycle would be as follows: Growth would be rapid from September to December, decreasing in January, February, with possibly a slight upswing before going into dormancy in May to August. Flowering would occur in October-November with fruit maturing in June to August.

Yield potential

Quoted yields vary greatly and apply to hand harvested blocks using standard cultural procedures. The range is from 1000 to 4700 kg per ha of dried beans. Plants take 2 to 3 years to come into production. Where stumping is practised, which is likely to be the procedure in Australia, one year out of four would be a non-production year. It could be expected that by using updated and mechanized production techniques and clonal material, above world average yields could be achieved.

Production of 2000 kg per ha per year would be required to make the industry viable. This appears feasible. Full mechanization of harvesting is needed however, and is possibly the major hurdle.

Market acceptability

The Australian market based on Australian Bureau of Statistics (A.B.S.) figures appears to be at least 28 000 tonnes of unroasted coffee for 1979-80. A further 190 tonnes of roasted coffee and coffee substitutes are also imported.

Figures for 1977-78 show sales of 1837 tonnes of ground coffee and 8321 tonnes of instant coffee indicating the bias to the latter. A.B.S. does not distinguish between Arabica and Robusta coffee. A reliable source has nominated that ground coffee is approximately 80% Arabica and 20% Robusta while the instant product exceeds 80% Robusta. An approximation of the market needs, using these figures would be 5600 tonnes of Arabica and 22 400 tonnes of Robusta.

Coffee grown at Buderim and Kuranda in Queensland at and about the turn of the century was considered then of good quality. There appears to be no reason why portion of the Australian market could not be satisfied with Australian grown coffee.

Using figures quoted above 2800 hectares would be required to supply the Arabica requirements alone.

Export potential

The world production of coffee is 4.6 million tonnes making Australia's consumption of 2800 tonnes insignificant.

World coffee marketing is under partial control of the International Coffee Agreement embracing both producing and consuming countries. The agreement aims to regulate trade by a system of export quotas. Quotas are brought into force when prices fall below a trigger price which was set in 1976 at $US 1.65 per kg. Since that time prices have not fallen below $US 2.64 per kg. Recent attempts by the International Coffee Organisation under the auspices of the U.N. Conference on Trade and Development (UNCTAD) have failed to obtain agreement on trade although a new trigger price of $US 2.95 per kg was generally agreed to although not implemented. A group of Central and South American producers, now known as the Bogota Group, applied a self-imposed ban on exports in 1978. Some stability has been brought to the market but further negotiations are proceeding.

From the above it is obvious the world market is well supplied and there appears no avenue for export from Australia. The one factor worthy of comment is the strong possibility of dumping or cutting of prices from overseas to undermine a developing local industry. In addition local processors having access to ready supplies from overseas could easily manipulate the local price and industry.

Cultural details

Propagation

The majority of coffee is produced from seed. With Arabica a high selfing rate produces fairly uniform types. In Robusta compatible parents are needed to maintain a good line.

Clonal material gives the normal advantages and cutting and grafting techniques are used. Cuttings using single node pieces are used hence a high rate of multiplication is possible. It appears that in many countries the benefit of clonal material is overshadowed by disease, pest and environmental limitations.

In vegetative propagation care must be taken to use the orthotrophic wood, as described previously, to get normal tree production. Of possible research interest would be the growing of meadow orchards using cuttings of the plagiotrophic branches particularly those of the Bourbon variety which grow more upright. In such orchards earlier cropping may be possible, the crop would be mown, stripped, and the wood used again for cuttings.

Rootstocks with nematode resistance are available which could be important where light soils, which are more prone to nematode activity, are used for crop management. Stock/scion compatibility however, requires investigation as problems are reported.

Management

Shading is generally considered as part of coffee culture but most coffee is now produced without shade. Shade is useful in areas of extreme cold and heat and also in lightly fertilized areas. Under Australian conditions site selection should overcome the need for shade as the economics of using shade are prohibitive and would impede mechanization.

Pruning has been widely used to control the crop, maintain bean size, facilitate harvesting and facilitate pest and disease control. Work has shown that free growth over the first 4-year production period has outyielded any form of pruning. After this time the tree becomes compacted and general yield and quality decline.

The apparent recent trend and that most likely to fit in with mechanization is one of a multiple stemmed plant which is regenerated by stumping every fourth year. Stumping infers the cutting back of the upright stems close to ground level to regenerate a new series of stems. This gives a cycle of 1 year of regrowth, followed by 3 years of production. The laterals would in fact carry, at the lowest point, 3 years of production by which time (or before) they would become too heavy. Quality can be maintained for this 3-year period on a lateral but declines after this time.

Some adaption of hedgerowing for mechanical harvesting would be required. Associated with this could be partial mechanical pruning of the lower laterals to maintain physical control of the size of the plant.

The management of pruning is greatly controlled by the vigour of the plant. Lack of vigour and extremes in the environment can cause the production of many upright and lateral secondary growths affecting the management and quality of the crop.

Density. Plant populations of 2500 plants per ha have outyielded densities down to half this population in the first two harvest years but fall to similar levels in the third and fourth years. Using the 4-year-cycle stumping technique above, populations approaching 2500 could be considered. The supply of clonally produced plants becomes a major factor in establishment costs.

With mechanization hedgerowing would be adopted and light inception studies would be a major consideration as it is known that self-shading towards the end of the above-mentioned stumping cycle depresses flower initiation.

Nutrition. Growth is apparently important in the first part of summer to maintain productivity. Flower initiation occurs in the autumn-winter period and fruit develop during the summer months. The management of nutrition would be required to service these demands. Reported nutrition work does not take these factors into concern.

Rates of up to 2 tonnes per ha of 12:4:20 (N:P:K) applied in four dressings per annum are nominated.

Irrigation

Irrigation will be imperative to the management, yield and quality considerations of growing coffee at high densities in lower rainfall light soil situations where crop management may be possible. It is obvious that the management of irrigation will need study as there appears to be little work reported overseas.

Pests and diseases

A number of major pests and diseases appear to attack coffee throughout the world. At present Australia appears to be free of any major problems.

Coffee leaf rust *(Hemileia vastatrix)* which has several races is the major problem. It occurs in Asia and the Pacific and has occurred in New Guinea where it was eradicated. Arabica coffee is susceptible although resistant cultivars have been developed. Robusta coffee is resistant and has replaced Arabica in some countries decimated by the disease.

Coffee berry disease *(Colletotrichum coffeanum)* causes serious losses in East Africa destroying the berries and causing dieback in the plant. 'Blue Mountain' a cultivar of Arabica shows resistance.

In each case both diseases can be controlled effectively by timely application of fungicides.

American leaf spot *(Mycena citricolor)* is the major disease in Central America. A range of other leaf spots occur. The severity of most of these leaf diseases is affected by the environment in which they occur.

Africa has the major pests as may be expected, being the seat of origin. Variegated coffee bug *(Antestiopsis orbitalis)* sucks young berries and stem terminals, causing significant damage and producing management problems. Other berry borers, stem borers, mealybugs, scale insects, leaf miners and nematodes attack coffee indicating that with the introduction of pests the costs of production could rise. Presently the incidence of pests is low in Australia.

New Guinea reports that the only disease they have of importance is pink disease *(Corticium salmonicolor)* which occurs only in wet areas. Pests include coffee ring and centre borers, *Meroleptis cinctor* and *Zeuzera* spp. respectively and also green coffee scale *Coccus viridus.*

Harvesting methods

Harvesting is the major variable cost in coffee production. The majority of coffee in the world is still hand harvested although several techniques have been adopted in an attempt to make the exercise more labour efficient.

A manual hand harvester is reported to pick 55 kg per day. In standard situations a planting requires three successive rounds to complete the harvest. In Robusta coffee the fruit is held when mature and does not fall, as with Arabica, so one round is sufficient.

The harvesting of the crop has been aided by the selection of varieties which tend to give even ripening characteristics. The use of ethephon (Ethrel) has improved this uniformity.

Ethephon applied at 100 ppm gives maximum ripening 6 to 13 days after spraying in Arabica and 9 to 17 days in Robusta. The use of gibberellic acid combined with ethephon may improve this result. Rates greater than 200 ppm have been reported to give loosening as well as ripening. Rates greater than 1000 ppm will give leaf drop and damage. Work has shown that the berries need to be at a certain physiological age and degree of uniformity to obtain good commercial results. It is obvious that cultivars of different origin will respond differently.

As coffee beans are on long supple branches normal shake harvesting is not effective. Beaters have been shown to remove the berries. It is reported that drum beater harvesters are under development in the Americas.

It is obvious that quality and the related price is imperative to the success of coffee production. Mechanical harvesting to be successful will require close management and manipulation of the crop to optimize quality. The technology of this management and an integrated harvesting system are still the major hurdles to the production of coffee in countries where labour costs are high.

Harvesters now used in blackberry production show potential for adaption to coffee. Multiple stem pruning, selection of cultivars and the use of crop control, and ripening and loosening regulators all show promise when considered in conjunction with beater harvesters.

Processing and storage characteristics

Harvested berries may be treated in two ways; either by drying and then hulling, or by pulping, fermentation, cleaning and drying. The former technique is used in Brazil while the latter appears to be more commonly used in other countries.

Where berries are pulped and fermented this must commence immediately as harvested berries start to ferment if left in bags or bins. Fermentation products are absorbed by the beans causing off flavours. Therefore, the process of fermentation needs to be completed within 48 hours to assure a high quality product.

Pulping, hulling, fermentation and drying plants should either be on a property or close by to offset the chance of off flavours in the product. Initial producers of coffee would have to allow for such facilities in their capital investment plan, possibly limiting pioneers to people with access to sufficient capital investment.

Marketing characteristics

Properly dried beans have a good storage life and they have a high value per unit weight. This makes coffee a viable crop in isolated areas where freight charges may be high and the transport service is irregular.

Chapter 25

RUTACEAE

WHITE SAPOTE by D.J. BATTEN
Casimiroa edulis Llave

Botany

Casimiroa edulis belongs to the subfamily Toddalioideae. It is a medium-sized, erect or spreading tree up to 15 m high (or occasionally higher) at maturity. Lenticels on young and old wood are conspicuous. Leaves are palmately compound with 3 to 7 leaflets. Leaflets are 70 to 130 mm long, lanceolate, sharply or sometimes bluntly pointed with entire or faintly crenate margins, glossy above, generally glabrous beneath. Flowers are small, greenish white, large in number, 5-merous, borne on panicles of between 15 to 100 or more flowers. Panicles are borne terminally on shoots with all leaves fully developed, in the axils of mature leaves, or on the bases of shoots a a year or more old. Fruits of selected varieties are 70 to 100 mm in diameter, spherical to oblate. The flesh is yellowish or white, melting, sweet and pleasant in the better varieties sometimes with a slightly resinous flavour. Skin is yellow to light green, thin and fragile. Seeds number 1 to 5, normally 2 or 3, large, enclosed in a hard, bony endocarp.

Climatic requirements

Distinctly subtropical in its climatic requirements, the white sapote does not succeed below 900 m elevation in Central America. In Mexico and Guatemala it is occasionally found growing at 3000 m.

Frost: The tree and its foliage are more resistant to frost than Guatemalan avocados, perhaps about as tolerant as oranges. Many light frosts will defoliate trees without other damage. Trees in California usually survive temperatures of –4°C.

Heat units: White sapote fruit quality is quite good in coastal areas of California which lack sufficient heat units for acceptable flavour development in citrus (Chandler 1958).

Drought: Although the tree will withstand drought, adequate soil moisture must be maintained for acceptable fruit size and yield to be obtained.

Soil requirements

Not much is known. According to Schroeder (1951) and Chandler (1958), the white sapote will not tolerate wet, poorly aerated soils, and nitrogen requirements appear to be less than that needed for orange trees.

Production season and time of cropping

There are two distinct types:

- Seasonal — main period of bloom in spring and crop matures mainly summer and autumn in California (for example, Dade, Pike).
- Everbearing — almost continuous flowering (for example, Wilson, Suebelle).

Genetic diversity

A number of varieties with widely differing characteristics have been named, mainly in California and Florida. A closely related species, the matasano *(Casimiroa tetrameria* Millsp.*)*, has velvety leaves, is larger and has an inferior flavour compared with *C. edulis.*

Some white sapote cultivars

Variety	Tree	Yield	Firmness	Flavour	Colour	Size (g)	Seeds	Other
Dade			moderate	excellent	yellow-green	230/ medium	4 or 5 medium	seasonal
Wilson			moderate	excellent	green			everbearing
Suebelle		moderate	fragile		green	small		everbearing
Pike	vigorous	prolific	very fragile	exceptional	light green	280/ large	3 or 4 medium	seasonal
Coleman	dwarfed			resinous	yellow-green	small	small	
Page Sarah Jones								
Golden		prolific			golden yellow			seasonal
Maltby Blumenthal Denzler Harvey Maechtlen								

Yield potential

Little information is available. Good varieties produce heavy crops. Some varieties have a pollen sterility problem and do not yield well (Mustard 1954).

Market acceptability

The primary outlet at present is as a fresh fruit. The sweetness of this fruit does not appeal to everyone. Some varieties have a bitterness which is objectionable but the better quality varieties, which are free of this characteristic, are liked by most who taste them. Sliced white sapote with cream is esteemed by many. White sapote could possibly be sold as a frozen product or as a flavouring for ice cream, yoghurt and confectionery goods.

Export potential

Markets would have to be developed. The perishability of the fruit would necessitate air-freight transport.

Cultural details

Propagation
White sapote does not come true-to-type from seed. Seed does not store well. Seed should not be allowed to dry thoroughly and should be germinated as soon as possible after removal from the fruit. Germination occurs in 3 or 4 weeks in warm weather or

with artificial heat (30°C) in cold weather. Seedlings grow quickly and are T-budded in spring or autumn when stem diameter is about 10 mm. Seedlings can also be grafted (whip graft) whenever it is warm enough for the trees to grow well. The scion piece needs to be wrapped or covered to prevent dehydration.

Last season's growth which has hardened off (grey bark) makes the best budwood. Older wood which has fruited will give poorer results.

Trees up to 150 mm in trunk diameter can be topworked by cutting off squarely and using bark grafts. The exposed cut and scion should be sealed against dehydration. Larger trees should be lopped, allowed to shoot, and the new growth budded or grafted.

It is not known whether white sapote can be propagated by root cuttings or air layering.

Management

Young trees need some training to avoid acute angles between branches. Tip-pruning will make the shoots branch and produce a good framework. Without pruning the growth will be excessively long and willowy and the mature trees will be prone to limb breakage. Infrequent heavy prunings will induce excessive suckering.

Grafted trees reach about 10 m in diameter so plant spacings should take account of this.

Irrigation

Young trees require frequent watering to encourage growth. Mature trees are drought tolerant but require adequate soil water during the period of fruit development to produce fruit of adequate size and quantity.

Pests and diseases

White sapote is rarely grown in commercial orchards so pest and disease problems have not been properly identified. Fruit fly affects white sapote badly. The everbearing types could be a particular problem in regard to fruit fly. Brown olive scale infests white sapotes in California but can be controlled simply by controlling the ants which farm them. Because the fruit ripens after picking, birds and fruit bats should not be a problem.

Harvesting methods

The fragile skin of white sapote necessitates hand harvesting from the tree.

Marketing characteristics

Most people who like sweet fruits find the white sapote very much to their liking.

Fruit picked at the mature-firm stage ripens in 3 to 5 days and is ready to eat when soft to touch. Some education of novice consumers would be needed to ensure fruit is allowed to ripen properly before being consumed. The yellow-skinned varieties have good eye-appeal but the green-skinned types are unlikely to attract many buyers. Large seeds are a drawback for the fruit.

Storage characteristics

The fruit ripens in 3 to 5 days at ambient temperatures. I have no information on the effects of refrigeration, ethylene absorbents, or controlled atmospheres on post-harvest life. Once the fruit softens it is very difficult to handle, so the prospects for marketing soft fruit (as has been proposed for avocados) would be minimal. Rotting is a problem and nothing has been published on controlling this problem.

Transport characteristics

Most varieties have a very thin skin and also bruise easily. Thomson (1972) says that even when fruit are handled carefully, rotting is a problem. The short post-harvest life at ambient temperatures means that transport must not take more than a day or two.

Processing characteristics

Jelly: When sufficient lemon juice is added to give white sapote jelly some character, it tastes like lemon jelly.

Frozen fruit: The fruit freezes well either whole or as pulp. When thawed it retains its original flavour.

Flavouring: White sapote apparently makes a delicious flavouring for dairy products such as milk shakes (Thomson 1972).

Medicinal value: In Mexico the fruit is believed to induce sleep (when sufficient is eaten) and also soothe rheumatic pains. The leaves, bark and seeds apparently contain a sedative glucoside. Tea made from the leaves of white sapote has been used by Spanish speaking people to relieve high blood pressure.

Wood: The wood is moderately hard, works well, and takes a smooth finish.

References

Chandler, W.H. (1958), 'Evergreen orchards'. Lea and Febiger, Philadelphia, p. 201.

Mustard, M.J. (1954), Pollen production and seed development in the white sapote. *Botanical Gazette*, **116**, 189.

Schroeder, C.A. (1951), The white sapote, *in* Texas Avocado Society Yearbook 1950, p. 60.

Thomson, P.H. (1972), The white sapote. California Rare Fruit Growers' Newsletter 4(3). (Also in California Rare Fruit Growers' Yearbook 1973 (vol. 5) — most recent paper on *C. edulis* abstracted in Horticultural Abstracts).

Further Reading

Biale, J.B., and Barcus, D.E. (1970), Respiratory patterns in tropical fruits of the Amazon Basin. *Tropical Science* **12**, 93.

PUMMELO
Citrus grandis (L.) Osbeck

by K. R. JORGENSEN

Botany

Citrus grandis belongs to the orange subfamily, Aurantioideae of the family Rutaceae. Synonyms are *C. maxima* (Burm.) Merril, and *C. decumana* (L.).

The preferred common name in English is pummelo which appears to be derived from pompelmoes or pomplemoose — the name used by the Dutch in the East Indies (Indonesia). The name 'pomelo' does not apply to this fruit but to the grapefruit, which appears to be a hybrid of the pummelo and some other citrus. Hodgson (1967) indicates that early in the present century the name pomelo was proposed and for a time used by American horticulturists for the grapefruit, but it was not accepted by the industry and has now virtually disappeared.

The name shaddock, used for the pummelo in the West Indies and the United States, derives from a Captain Shaddock, commander of an East India ship who is reputed to have introduced the plant to Barbados as seed in the late 1600s (Hodgson 1967).

Names used in other languages and countries include:
French – pamplemousse; Italian – pompelmo; German – pampelmuse; Spanish – toronja; Japanese – buntan or zabon; Indonesian – jeruk besar, limau besar or jeruk gulung; Malay – limau Betawi (Batavian lemon) or limau Bali (Bali lemon); Thai – Som-o or Ma-o; Hindi – batabi, chokotra; Chinese – wentan.

The origin of the pummelo is difficult to establish precisely. Because many wild pummelo seedlings are widely distributed in Indonesia and Malaysia, some authors consider it to be native to this region. However, there is also evidence that the pummelo and many other citrus species were taken throughout South East Asia by Chinese settlers and that the pummelo has naturalized in these regions (Martin and Cooper 1977).

The pummelo was mentioned in Chinese literature in 2200 B.C. It was mentioned as Adam's apple when seen in Palestine in 1187 A.D. and in Spain about the same time. It was described in Jamaica in 1696 under the name shaddock. It is now cultivated in tropical and subtropical regions throughout the world.

The pummelo is a typical citrus with shiny dark green leaves and large fruit containing juice-filled pulp vesicles. It is one of the most distinct species of the *Citrus* genus and can be separated from other citrus species by its young shoots which are pubescent (covered in fine hairs), its huge leaves with very broad wings on the petiole and its very large fruit.

The flowers are large, 3 to 7 cm in diameter, and white or cream in colour. The fruit are also large, 10 to 30 cm in diameter and may be either globose, slightly flattened or pear-shaped depending on the variety. The apex may be round or slightly depressed and the style and stigma may persist on the fruit and grow with it. At maturity the fruit is light green to yellow.

The rind is soft, pithy, pink or white and from 1 to 3 cm thick. The surface of some fruit is downy. The centre of the fruit may be solid but more commonly is open. The segment walls of the fruit are tough but can be easily separated from the large juice vesicles without bursting them and spilling the juice.

The pummelo has some similarities with the grapefruit. However the grapefruit is not pubescent on shoots, leaves or fruit and its seeds are polyembryonic (several embryos in one seed) whereas the pummelo is always monoembryonic.

Climatic requirements

The pummelo is best suited to a tropical climate although some varieties produce satisfactory fruit in the subtropics also. The tree grows satisfactorily in a temperate climate but the fruit produced is usually excessively acid.

According to Yee (1974), the best fruit in Hawaii are produced in warm locations below 150 m elevation with a temperature low of 15 to 18°C and a high of 29 to 32°C. Pummelos have considerably less cold tolerance than grapefruit, although their heat tolerance is similar (Hodgson 1967).

Soil requirements

The pummelo, like other citrus fruits, grows best in deep well-drained sandy loams of moderate acidity. Poorly drained soils are likely to produce root rot problems.

The pummelos of highest repute in South East Asia are grown in reclaimed coastal marshes in Thailand, where they are subject to the flow of brackish tidal water. This has given rise to the belief that the pummelo is salt tolerant and that salt improves fruit quality. Investigations by Monselise (1960) and Cooper and Gorton (1952) indicate that the pummelos tested were not specifically salt tolerant.

Genetic diversity

Because the pummelo is monoembryonic, it does not come true to type from seed (as do the majority of citrus varieties which are polyembryonic). As well, most pummelos are self-sterile or nearly so (Soost 1964) so that most seeds are produced by cross-pollination and the resulting seedlings are variable. Very few seedlings produce fruit of acceptable eating quality.

Unfortunately there are severe quarantine restrictions on the importation of all citrus budwood into Australia in order to prevent the introduction of new virus and mycoplasma diseases. This currently precludes the importation of superior pummelo varieties from overseas. As a result there are few good pummelo varieties available in Australia.

A few satisfactory pummelos have been located in north Queensland recently but they are not yet widely cultivated. Ryan is an early season (April-June), white fleshed variety, Bosred is mid to late season with pink flesh, while Termat is mid-season and white fleshed. These varieties are now being propagated commercially. Three selections from New South Wales (accession numbers 3589, 3238 and 3129) have been tested at Nambour but they produced unpalatable acid fruit with thick rinds.

Many good pummelo varieties are available overseas. Some have white flesh, others pink. Completely acidless varieties are available but these are not popular with Europeans.

Of the varieties grown in the United States, Martin and Cooper (1977) consider the following have the most promise: Red Shaddock, Thong Dee (seedling line), Webber, Chandler, Kao Panne, Siamese, Reinking and Nakon. According to Nauer *et al.* (1976) Chandler is better adapted to subtropical climates than most South East Asian varieties.

The most famous cultivar in China is the Shatien.

Thailand has a very good gene pool and the most popular cultivars are Khao Phuang, Khao Thongdee, Khao Panne, Khao Yai and Khao Jeep.

N.B. *Propagating material must never be imported illegally from South East Asia as most of the countries have citrus canker, Asian greening and severe strains of tristeza virus which would severely damage the Australian citrus industry if introduced here.*

Production season and time of cropping

In a hot tropical climate, pummelos bloom several times a year but in Australia the main flush of flowering is in spring following the dormancy induced by cold and low moisture in winter. The fruit matures 6 to 9 months later.

Yield potential

No details of yield are available but the pummelo is expected to crop as well as other citrus with a yield of at least 100 kg per year from mature trees.

Market acceptability

The flavour of good quality pummelo is pleasantly sweet and mildly acidic with a slight hint of bitterness. The thick rind and tough segment walls however make it difficult to serve the fruit like a grapefruit. Instead the fruit is cut open and the segment walls stripped from the pulp sections. Fortunately the pulp vesicles separate easily without breaking and can be served and eaten without the juice escaping. It makes a pleasant addition to a dessert or salad, and is also popular as a breakfast fruit.

The size of the fruit and its thick rind are a disadvantage for marketing because of the extra cost involved in packaging and transport. Small fruited varieties are therefore preferred. Its main advantage is its low perishability. The fruit do not bruise easily and can be stored for up to 2 months under suitable conditions. The pummelo is expected to be a specialty fruit rather than one that sells in large volumes.

Export potential

The large size of the fruit makes it difficult to export economically. In the South East Asian markets where pummelos are already well known, a limited market may exist at the Chinese New Year.

Cultural details

Propagation

Seedlings are not recommended because they are unlikely to produce worthwhile fruit, Marcots produce satisfactory trees and this is the main method of propagation in South East Asia

Budded trees are, however, the most convenient for Australian conditions using the standard T method of budding as for other citrus. Rough lemon has been used successfully as the stock but other stocks are likely to be useful and will produce better fruit quality. Martin and Cooper (1977) recommend pummelo seedlings as stock.

Management

The trees grow quite large and an 8 x 8 m spacing is needed. The fruit will be less seedy if the trees are grown in isolation from other citrus varieties.

Young trees should be pruned to establish three equally spaced scaffold limbs. Further pruning is aimed at removing dead branches or those which interfere with mowing and spraying.

Irrigation

Irrigation at regular intervals of 7 to 14 days will be needed to keep the trees adequately supplied with moisture during dry spring weather.

Competition for moisture can be reduced by keeping the area under the tree free of weeds with a knockdown herbicide such as glyphosate. Interrow spaces should be covered with a mown sod to prevent erosion.

Fertilization

Fertilizer applications should be sufficient to supply the tree annually with 100 g nitrogen, 10 g phosphorus and 50 g potassium per year of age. Copper, zinc and manganese should be applied by spray when the trees show symptoms of deficiency. Soil pH should be maintained in the 6 to 7 range by the application of dolomite when necessary.

Pests and diseases

Pummelos are subject to the same pests and diseases as other citrus. Scale insects can be controlled when necessary with oil sprays and fruit fly with maldison bait sprays. Root and collar rot problems can be reduced by using well-drained soils and keeping the trunk free of weeds and mulch.

Harvesting methods

Fruit can be left on the tree for several months after it colours. It is attacked by fruit fly but this does not cause the fruit to fall perhaps because the thick rind prevents maggot development.

In Thailand, the fruit are picked just as they are starting to lose their dark green colour. They are then stored for 1 to 2 months during which time the fruit improves in juiciness and flavour (Martin and Cooper 1977).

Marketing, storage and transport characteristics

The fruit is easy to market as it stores well and does not bruise easily. A benomyl dip could be used to prevent breakdown due to blue or green mould. Its large size makes it expensive to pack and market.

Processing characteristics

The fruit is not commonly processed but would be suitable for canning in a similar way to mandarin segments.

In Asia, a very acceptable sweet is prepared from the rind.

References

Cooper, W.C. and Gorton, B.S. (1952), Toxicity and accumulation of chloride salts in citrus on various rootstocks. *Proceedings American Society for Horticultural Science*, **59**, 143.

Hodgson, R.W. (1967), Horticultural varieties of citrus, *in* W. Reuther *et al.* (eds), 'The citrus industry', vol. 1, University of California Press, p. 431.

Martin, F.W. and W.C. Cooper (1977), Cultivation of neglected tropical fruits with promise. Part 3. The pummelo. Agricultural Research Service, U.S.D.A.

Monselise, S.P. (1960), On the presumed salt tolerance of Siamese pummelos. *Proceedings of the Florida State Horticultural Society* **73**, 19.

Nauer, E.M. *et al.* (1976), Grapefruit and pummelos: C.V.I.P. virus-free budwood sources. *Citrograph* **61**, 363.

Soost, R.K. (1964), Self incompatibility in *Citrus grandis* (Linn.) Osbeck, *Proceedings American Society for Horticultural Science* **84**, 137.

Yee, W.Y.J. (1974), The pummelo in Hawaii, Cooperative Extension Service Circular 483, University of Hawaii.

Further reading

Swingle, W.T. and P.C. Reece (1967), The botany of citrus and its wild relatives, *in* W. Ruether *et al.* (eds), 'The citrus industry', vol. 1, University of California Press, p. 190.

Chapter 26

SAPINDACEAE

LYCHEE
Litchi chinensis Sonn.

by K.R. CHAPMAN

> Beneath these green mountains where Spring rules the year
> The arbutus and loquat in season appear
> And feasting on lychee — three hundred a day
> I should not mind staying eternally here.

Thus sang a Chinese poet 900 years ago about the luscious lychee native to southern China.

Botany

Groff (1944) considered that little is known of the lychee *(Litchi chinensis* Sonn.*)* except in its distant relationships. He records that *L. chinensis* has never been found in the wild, and that it is closely related to the numerous *Nephelium* spp. for which the centre of distribution was Malaya. He reported that the lychee we know today developed in the southern China region out of the *Nephelium* spp. via distribution northward from Malaya.

It is of interest to note that Hsu *et al.* (1964) reported the discovery of wild lychee trees, so perhaps the picture is not as clear-cut as Groff (1944) would have us believe.

The litchi genus contains one other member, *Litchi philippinensis,* which is not known in Asia and seems to have developed in the Philippines as a result of altitude, while *L. chinensis* evolved in China as a result of latitude and careful selection by the Chinese.

Storey (1973) noted that the lychee is known to have been cultivated in southern China as early as 1766 B.C.

The literature contains many detailed accounts on the botanical descriptions of lychee, its floral biology and growth. The description which follows is from Storey (1973) who gives a concise account.

The botanical name of the lychee is *Litchi chinensis* Sonn. The species is a member of Sapindaceae, the soapberry family, which comprises about 2000 species of tropical and subtropical trees, shrubs, herbs, and vines that have been classified into 140 genera. Some of the species which are cultivated in California as ornamentals are: *Dodonaea eriocarpa,* a colourful shrub often used for hedging; *Filicum decipiens,* the fern tree; *Harpullia pendula,* the tulipwood tree; *Kolreuteria formosana,* the golden-rain tree; and *Cupania anacardiodes,* the tuckeroo. Species grown elsewhere in the world for their edible fruits are: *Euphoria longan,* the lungan (longan) or dragon's eye, a few trees of which are to be seen in California; *Melicoccus bijugatus,* the Spanish lime of Florida and the Caribbean region; *Blighia sapida,* the akee, which is the national tree of Jamaica; and *Nephelium mutabile,* the pulasan, and *Nephelium lappaceum* L., the rambutan, both of the Malaysian-Indonesian region. A related species native to California is the California buckeye, *Aesculus californica.*

The lychee tree is a medium to large, handsome evergreen with a short, stocky trunk and a low, large head. In Florida, Hawaii and elsewhere, trees reach a height of 14 m, with an approximately equal spread when grown under favourable conditions. In some varieties, the branches are crooked or twisting, and low-hanging and spreading forming a head broader than high. In other varieties, the branches are fairly straight and more or less upright, forming a compact, rounded head.

The leaves of the lychee are arranged alternately on the stems. They are pinnately compound, consisting of a leaf stalk (petiole) and its extension (rachis) with 2 to 5 pairs of leaflets. The combined length of petiole and rachis is 7 to 28 cm. The leaflets are arranged in opposite or slightly oblique order along the rachis on short stalks (petiolules) of their own which measure 3 to 25 cm in length. The blades of the leaflets are a lustrous deep green on the upper surface, and a waxy lighter green (glaucous) on the under surface. They are soft-leathery to the touch. Mature leaflets measure 7.5 to 20 cm in length and 2.5 to 6 cm in width. They are elliptical to lance-shaped. The bases of the blades are wedge-shaped, or even more or less rounded. Length of petiole and of rachis, arrangement of leaflets on the rachis, number of leaflet pairs, and leaflet shape and dimensions are useful characters for identifying varieties.

The inflorescence is a many branched panicle which, depending on variety, may vary in length from 7.5 cm to over 30 cm. It bears hundreds of small white or yellowish flowers which give off a mild, unpleasant odor when the tree is in full bloom. The numbers of fruit maturing on a panicle may vary from 1 to 40, or more.

Lychee flowers measure about 3 mm to 6 mm across when fully expanded, resting on flower stalks (pedicels) about 1.6 mm long. They have 4 or 5 short dentate sepals, but are without petals. They have 6 to 10 stamens.

Five different types of flowers have been described in the literature on the lychee. Classification of the types is based primarily on length and functionability of the stamens, and on development and functionability of the pistil. The flowers are unisexual in function, so they may be classified broadly as male and female. Both classes of flowers are borne in the same panicle, but tend to alternate in their periods of opening.

The typical male flower has a vestigal or abortive pistil at its centre. Surounding this are the 6 to 10 stamens with filaments up to 6 mm long. A single flower remains functional for several days during which the pollen sacs ripen and shed pollen successively rather than simultaneously.

The typical female flower has a small, but fully developed, pistil which rests on a short stalk or gynophore. The ovary is two-lobed with each lobe containing an ovule. It is surmounted by a short style, which is cleft at the apex into two rays that expose white, sticky, stigmatic surfaces when expanded. The pistil is surrounded by the stamens which, however, have very short filaments, usually no more than 1.5 mm long. The pollen sacs do not open, so the flower sheds no pollen.

Generally, only one lobe of the ovary develops into a fruit, the other lobe aborting. Occasionally, both lobes develop, and the mature fruit superficially resembles two fruits adhering to each other at their bases, each containing a seed.

The typical one-seeded fruit may be round, ovoid, or heart-shaped, depending a good deal on the variety. The outer covering is a thin, leathery shell (pericarp), which is bright red in most varieties. It is rough in appearance and to the touch, its surface being covered with angular or conical protuberances. The shell becomes brown and brittle upon drying.

The fleshy edible portion of the fruit is an interesting structure. It is an outgrowth of the seed stalk, and is called an aril. It grows as the fruit develops until it completely envelopes the seed. It is white and translucent, and has a consistency like that of a grape. The flavour, which is subacid and quite distinctive in its own right, is likened by many persons to that of a prime muscat grape.

Upon drying, the aril shrinks away from the shell, and becomes shrivelled, brown, and date-like. It is in this form only that many persons were previously acquainted with the fruit as the 'Chinese lychee nut'.

The chestnut-brown seeds are ovoid to oblong, 10 to 23 mm long and 6 to 12 mm wide. In some varieties, a high percentage of the seeds in fully developed fruits may be abortive. The abortive seeds are small and shrivelled. The Chinese people call the abortive seeds 'Chicken tongue' seeds. Fruits with abortive seeds are preferred, for they are no smaller than fruits with normal seeds, and consequently, contain a larger proportion of flesh.

Climatic requirements

Goto (1960) reported that by the end of the 17th century the lychee was introduced to Burma and India from China, and to the West Indies by 1775. It was introduced to South Africa in 1870 and to Hawaii in 1873. Storey (1973) records that the lychee spread around the tropical and subtropical world in the 16th and 17th centuries and is thought to have been introduced to Florida in 1870 and to California in 1897.

The most important production areas today are China, Taiwan, India, South Africa, Thailand, Burma, Hong Kong, Macau and Mauritius. It will fruit in Florida, Hawaii, West Indies, Brazil, California and Australia. In Taiwan, lychees are produced in central Taiwan, where rainfall (amount and distribution) and temperature conditions are similar to those of southern Queensland and northern New South Wales. The areas are not far from the sea and are at an elevation of around 50 m (Gordon Vallance, personal communication).

Temperature and humidity

The climatic requirements and influences feature prominently in the lychee literature. Groff (1921) states that periodic cold between 30 and 40°F (–1°C and 4°C) in winter causes the physiological changes in lychee trees required for fruit bearing.

Groff (1944) stated that the lychee does not fruit successfully to the south of southern China because of warmth and to the north of this area because freezing temperatures limit its survival.

He also noted that the lychee is distinctly heat-loving and yet relatively cool winter temperatures are necessary for fruit bearing. In the Canton delta where lychees flourish and crop extremely well, air temperatures rarely drop below 32°F (0°C). Where winter temperatures never go below 50°F (10°C) the area is not conducive to fruit bearing, even though the tree will make beautiful vegetative growth.

Young lychees are killed or severely damaged at –2 to –3°C, and old trees are killed or severely damaged at air temperatures of –4 to –5°C (Storey 1973). Joubert (1970) reported that temperatures below 0°C damage lychee flowers.

Groff (1944) stated that in active growth the lychee thrives in a high humidity atmosphere, and in Canton the summers are noted for high humidity which is never below 80% and frequently above 90% both night and day. High rainfall and high humidity are conducive to maximum growth especially when combined with high temperatures in the ranges 78 to 91°F (26 to 33°C) maximum and 64 to 77°F (18 to 25°C) minimum (Groff 1944). If temperatures are too low (5 to 8°C) at anthesis, the flowers do not open.

Fruits do not split open in a very moist atmosphere, but they tend to dehisce (a character inherent in the family) in dry areas experiencing hot dry winds (Groff 1944). Indian varieties have been selected to tolerate the hot drier conditions. In India, very high temperatures (40 to 45°C) are experienced around maturity of the fruits (Gordon Vallance, personal communication). Joubert (1970) in South Africa considered that later splitting is reduced by keeping up soil moisture when the skin of the fruit is actively being produced by cell division during the first 60 days after pollination.

Rainfall, irrigation and wind

When rainfall is not ample (around 1270 to 1524 mm well distributed throughout the year (Storey 1973)) irrigation is essential (Groff 1944).

The distribution of rainfall is of more significance than the amount. A dry autumn-winter period is considered important for vegetative dormancy in lychee, so essential for good flowering (Nakata and Watanabe 1966).

Anthesis in southern China corresponds to the rainy season (130 to 375 mm per month) and relative humidities greater than 50%, and Storey (1973) points out that moisture in the air is necessary for pollen germination and setting. *This is not the case in Queensland where blossoming corresponds with dry spring weather.* However, too much rain at anthesis reduces flower opening and hence insect activity, particularly that of honey bees needed to pollinate flowers. Too little rain results in flowers that dry out and fail to set.

Wind protection is essential, both from the viewpoint of tree damage and with respect to cold dry or hot dry winds at setting, and near maturity with respect to fruit splitting. In Australia, good wind-breaks could also be expected to contribute to higher humidities in the orchard.

Light intensity and photoperiod

Young lychee trees show unfavourable reaction to intense light, demonstrating their forest origin (Groff 1944). Protection of young trees in the field from both wind and intense light is essential for at least the first 12 months.

Nakata and Watanabe (1966) demonstrated that the lychee is a day-neutral plant. The only effect of photoperiod was that short photoperiod increased the period of female flower anthesis, particularly when combined with longer cold periods. These workers showed also, that increased starch level in leaves and stems accompanied flowering while sugar levels stayed static.

Soil requirements

Groff (1944) recorded that in south China the soils favourable to lychee are either alluvial deposits of streams, as in the Canton delta, or the laterite soils of the hills, neither of which drains readily. He also noted that they thrive in organic soils (Chinese gardens), and in Florida on lighter sandy soils and meuck soils, but often seem to thrive best where sandy-clay subsoils exist. Groff also pointed out that the lychee will grow along the dykes of the Canton delta and will withstand 10 to 14 days of immersion, while Storey (1973) states they will not withstand immersion for long if the water becomes stagnant.

In Thailand, lychees grow on banks with high water-tables beneath them, while in Taiwan lychees are grown on hillsides on gravelly, shaley soils.

The favoured pH range is 5.0 to 5.5, and where pH is 6.0 nutrient deficiencies result and extra attention to fertilizing is necessary. However, Storey (1973) reports that in India and Florida, lychees will grow satisfactorily in neutral to slightly alkaline soils with free lime present. A statement by Colville in Groff's (1921) book points out that acid soils favour mycorrhizal fungal associations with lychee roots. These associations are reported to be beneficial and essential to growth and productivity.

The lychee is apparently more plastic with respect to soils than to climate. However, it has been our personal experience that in south Queensland, tree vigour is difficult to control on fertile soils, particularly on warm slopes.

It is a personal feeling that the lychee could be better managed (with respect to water stress in autumn-winter) on 'hungry' lighter soils, in cooler sites where nutrient stress is also easier to achieve for vigour control and hence promotion of flowering and cropping.

In summary, a north to north-east aspect with lighter soils that can be stressed readily for nutrients and water is preferred. Wind protection from the north-west and south-west is essential to avoid cold or hot dry winds, which upset fruit set, fruit retention and cause splitting. Wind protection on the south-east is also essential to minimize cyclonic wind damage. A cool site with light ground frosts is also required, and irrigation is essential for establishment and management. Cool dry autumn-winter conditions are needed.

Genetic diversity

Groff (1921) has listed 74 varieties of lychee of Chinese origin, many of which are grown widely today. Singh and Singh (1954) report 33 varieties grown in India all of local origin derived from seeds of Chinese varieties introduced into India in the late 18th century.

South Africa essentially bases its industry on one lychee variety H.L.H. Mauritius, while in Hawaii, a local seedling of Haak Yip called Groff is reputed to perform better than the traditional Chinese varieties. Taiwan bases 90% of its production on Haak Yip (Don Batten, New South Wales Department of Agriculture, personal communication), while India uses its local varieties for production.

There are many descriptions of lychee varieties in the literature, but Groff (1921) in his famous book 'The lychee and longan' and Singh and Singh (1954) cover the most prominent varieties of China and India respectively.

Groff (1921) pointed out that the Chinese recognize the extreme difficulty in perpetuating desirable characteristics of highly prized varieties under conditions other than those in which the fruit had its origin — a point which suggests heavy selection pressure for centuries on the lychee, and of great importance to us in growing these varieties under Australian conditions.

He said that the cultural requirements for different varieties are recognized by lychee specialists in China who recommend to the average farmer the varieties which may be best adapted to his conditions. He also noted the two general classifications of lychees namely the 'mountain' and 'water' types with the latter containing most of the better types.

The following is a list of the more commonly grown varieties in Kwangtung province of southern China (Groff 1921). The first name is the Mandarin and the name in brackets is the Cantonese equivalent. (N.B. The pinyin, and cantonese names now proposed for Australian adoption vary from Groff's nomenclature.)

(1) No mi tsz (No mai tsz) — glutinous rice
(2) Kua lu (Kwa luk) — hanging green
(3) Kwi wei (Kwai mi) — cinnamon flavour
(4) Hsiang li (Heung lai — fragrant lychee
(5) Hsi chio tsu (Sai kok tsz) — rhinoceros horn
(6) Hei yeh (Hak Ip) — black leaf
(7) Fei tsu hsiao (Fi tsz siu) — imperial concubine's laugh
(8) T'ang Po (T'ong pok) — pond embankment
(9) Shang shou huai (Sheung shu wai) — 'President of the board embraces'
(10) Ch'u ma tsu (Chu ma tsz) — Chinese grass fibre
(11) Ta tsao (Tai tso) — large crop
(12) Huai chih (Wai chi) — the Wai River lychee
(13) San yueh hung (Sam ut hung) — 'third month red'
(14) Pai la li chih (Pak lap lai chi) — white wax litchi
(15) Shan chih (Shan chi) — mountain lychee, considered of poor quality and of use only as a rootstock. Mountain lychees usually have irregular trunks with buttresses. The Taiwanese regard the Brewster as a mountain lychee type (Gordon Vallance, personal communication).

Groff (1921) also mentions others of importance, and a more recent publication in Chinese (interpretation provided by Colin Bock, Bundaberg) mentions Souey tung as a variety grown in China in recent years.

Chadha (1968) lists the following varieties of importance in India. Bihar State: Desi, Purbi, China (early maturing), Kasava, Bedana, Early Bedana, Dehra Rose, Shahi (mid-season), and Late Bedana (late maturing). Some of these are also grown in Bengal State. Uttar Pradesh State: Early Large Red, Early Bedana (early maturing), Late Large Red, Rose Scented (mid-season), Late Bedana and Kalkattia (late season). Others grown in Uttar Pradesh include Extra Early Green, Gulabi, Pickling, Khatti, Dehra Dun, and Piazi.

Sa Kan is a new large fruited variety in Taiwan that is apparently of good quality (Gordon Vallance, personal communication).

Some important indian lychee cultivars. Prepared by B. J. Batten

Cultivar	Synonyms	Season*	Yield	Vigour	Average fruit weight (g)	Seed
Saharanpur	Early Large Red	early	moderate to high	vigorous	22	large
Dehra Dun		early	very high	medium	18	large
Muzaffarpur	Shahi, Late Large Red	early/mid	very high	vigorous	20	large
Rose Scented		mid	moderate	medium	16	large
Seedless Late	Late Seedless, Late Bedara	late	low to moderate	very vigorous	29	small
Calcuttia Late	Calcuttia, Kalkattia, Calcutta	late	very high	weak	20	large

* The season ('early' to 'late') is 5 weeks long.

NOTE: This information is based on Indian information and cultivars may not be available in Australia.

In Australia, we have a number of lychee varieties and selections but the two most commonly grown to date are Tai so and Bengal. The Kwai Mi, so called in Queensland, does not fit the Chinese descriptions of this variety. The variety presently known as Kwai Mi in Queensland is not the Chinese Kwai May but Tai so. Bosworth No. 3 in Queensland is the Chinese Kwai May – Pink (Gordon Vallance, personal communication). Other varieties now gaining importance are Haak Yip, Wai Chee, Gee Kee and No Mai Chee.

There is a deal of confusion about varieties and varietal types in Australia at present and these need to be carefully examined and appropriate names given. As a suggestion where Chinese varieties are concerned, we should accept the Cantonese name and not the Mandarin one, since Canton is the area closely allied with its earlier origin and greater production.

In summary, the known genetic diversity is wide, but with any particular variety the adaptation would seem to be limiting, particularly with the heavily selected Chinese varieties.

Production season and time of cropping

The production season in Queensland extends from the first week of November in north Queensland to mid-March in southern Queensland at Mt Tamborine. In China the extent of the season is from the third month (early May) for Sum yee hong ('third month red') to the seventh month (September) for the very latest maturing varieties, for example, Wai chee.

Some important Chinese lychee cultivars (Guangdong Province)

Prepared by D.J. Batten

Cultivar (Pinyin)	Cantonese name (Aust. system)	Meaning of name	Season*	Yield	Vigour	Fruit quality	Average fruit wt (g)	Proportion edible %	Seed	Use	Other comments
San yue hong	Sum Yee Hong	Third month red	early-mid May	regular	medium	fair	40	62–68	large	fresh consumption	Earliest cultivar
Shui dong	Souey Tung	Water east	late May/ early June	regular	dwarf	good	25	72	large	fresh	Fruit similar to Hei yue
Da zao**	Tai So	Big Crop	early/mid June	regular	vigorous	fair	24	72–77	large	fresh	Susceptible to wind damage
Hei yue	Haak Yip	Black Leaf	mid-June	high, regular	moderate	good	19	78	large	fresh, canned	Susceptible to wind damage
Fe Zi Xiao	Fay Zee Siu	Concubine laughing	early/mid June	low, irregular	vigorous	excellent	30	79–83	small	fresh	Historically rather than commercially important.
Gui wei	Kwai May	Sweet-scented Osmanthus aroma	late June/ early July	medium, irregular	vigorous,	very good	17	70–83	mostly small	fresh	Sweet long before full maturity; 'thorny' skin.
No mi ci	No Mai Chee	Glutinous rice cake	late June/ early July	medium, irregular	vigorous	excellent	25	82–86	small	fresh, dried, canned	A prized variety.
Shang shu huai	Seong Sue Wai	Minister's bosom	late June/ early July	moderate	vigorous	very good	17	73–77	large & small	fresh	
Huai zhi	Wai chee	Bosom litchi	early/mid July	high, regular	dwarf	good	21	72	mostly large	fresh, dried, canned	Latest cultivar, wide adaptation

* At Guangzhou, China: varieties listed in normal order of maturation.
**Synonyms: Hong huey, Maw mong (Thailand), H.L.H. Mauritius (South Africa), Kwai mi (U.S.A.), Australia — incorrect name).

NOTE: This information is based on Chinese information and cultivars are not all yet available in Australia.

Varieties are needed in Australia to extend the production season to each end as it is still very short. Longer market exposure is desirable for promotion of the industry.

Lychees will crop at around year 3 in the field, but do not seem to produce heavy crops until around year 7 to 8, and go on increasing in yields up to 20 to 30 years of age.

The lychee is a long-lived tree growing up to 20 m high with a girth of 4 m.

Yield potential

The lychee is a heavy bearer when conditions favour fruiting. Seven and eight-year-old trees should produce up to 45 kg of fruit, with 10 to 12-year-old trees carrying 60 to 70 kg. Fifteen to twenty-year-old trees produce 130 to 140 kg and still older trees should produce in excess of 150 kg and up to 300 kg. Yields are variable in Queensland and northern New South Wales, and in some instances biennial in nature. Flowering, fruit set and fruit retention problems are apparent, particularly with younger trees 5 to 10 years old, which are often excessively vegetative.

Market acceptability

'It is the most tasty and beautiful fruit that God created in the Universe', and 'One of the daintiest packages that has ever been wrapped by Nature's hand. No-one, whether he is a student of nature or not, can examine the litchi without admiring its beauty and the sanitary method by which it is preserved'. Thus remarked the early travellers to China while extolling the virtues of lychees (Storey 1973).

Market acceptability of lychees is high but will vary as the fruits become more popular and plentiful because of varied varietal characters, particularly seed size and richness of flavour, as well as general appearance. In Hong Kong the No Mai Chee from mainland China attracts about twice the price as Kwai May, and about 5 times the price as Haak Yip from Taiwan (Gordon Vallance, personal communication). Storey *et al.* (1953) consider the flavour of Groff variety to be delicious and only surpassed by the Lokong clone of No Mai Chee and the Punyu Tuwah Suitung clone of Haak Yip.

Production for the wet tropics in Queensland was 35 000 x 5 kg cartons in 1979. At present this area would account for 90 to 95% of Queensland production.

Export potential

Because Australia grows lychees out of season with respect to the Northern Hemisphere, the export potential for lychees grown in Australia is good. Mauritius and South Africa cater to very much under-supplied European markets and the potential for marketing throughout South East Asia and Japan is high. Fresh lychees could be landed on any of these markets by air freight. The South East Asian markets will be very discerning with respect to fruit quality.

Speed is the essence of the exercise to preserve the bright attractive appearance of the fruit. Precooling, cool storage, freezing and drying also add to flexibility in export market development with lychee (see later sections).

Cultural details

Propagation

Marcottage or aerial layering is the most popular technique for propagation of lychee and the success rate is high.

Cuttings and budding and grafting can also be used to successfully propagate lychee. These techniques are more economical with respect to scion wood, but cuttings and budded or grafted plants seem to take longer to establish before field planting and the success rate of take is lower.

In our experience, we have had a number of graft union collapses in the field, due to incompatability while plants were in their first or second year.

In certain areas of China approach or inarch grafting of seedlings is used with the No mai chee scion. The seedling rootstock used is that of the 'mountain lychee', Shan Chee. It is thought to impart superior attributes to the already superb fruit quality of the 'chicken tongued' (small aborted seed) fruit of the No mai chee. In Hong Kong, lychees are approach grafted onto Wai chee seedling rootstocks (Gordon Vallance, personal communication).

Rootstocks may have a future for use in lychee, with respect to vigour manipulation or perhaps disease or pest control. However, at present, marcots are the preferred method of propagation followed by cuttings. For a good account of marcotting techniques see Cull and Hams (1974).

Management

Pre-plant

(a) Deep ripping and pH adjustment up to 5.0 to 5.5.
(b) Organic matter addition to planting sites, for example, mill-mud (35 L per m^2) or fowl manure (10 L per m^2).
(c) Fumigation for 'ring' nematode, *(Hemicriconemoides manganifera)*, 'dagger' nematode, *(Ziphinema brevicolle)* and 'root knot' nematode *(Meloidogyne javanica)* where they are known to occur, plus a soil mite *Sarcoptiformes* spp., all of which contribute to lychee decline.
(d) Addition of single superphosphate 0.5 kg to planting sites (1 m^2) dug in pre-plant.

Planting distances: 12 x 12 m, 70 plants per hectare for Kwai May, Bengal, Mauritius, Haak Yip and 10 x 10 m, 136 plants per hectare for Wai Chee and No Mai Chee. High density planting to be thinned at plant intersection, 12 x 6 m and 10 x 5 m respectively for the two varietal groups.

Planting

Direct field planting of marcots is not recommended. Marcots should be grown in pots or polybags for 2 to 3 flushes (6 to 9 months) hardened off and planted preferably in the wet season. On colder sites, spring planting can be considered.

When marcots are removed from trees half of the foliage can be removed. In particular, remove side shoots that will produce narrow crotch angles with varieties such as Tai So, Kwai May, Bengal, and Haak Yip. When planting or when potting marcots handle the roots carefully as they are brittle and subject to damage.

When planted in the field always use a hessian or Sarlon tree guard to surround the plant (0.8 x 0.8 x 1.0 m high) for wind and intense sunlight protection.

Make sure irrigation is available to keep plants moist in this critical stage of establishment. Mulching around the plants is desirable and weeds should be kept well controlled along the tree line. In between the rows, a mown sward or clean cultivation is recommended.

With young trees up to 3 or 4 years of age, heading back of main terminals in spring (just prior to a flush) in the more vigorous cultivars, helps create a more compact tree with many terminals and thus potential bearing surface. Up to 15 cm of tip is removed on each occasion. On average, 3 new terminals are produced at each cut. Removal of low side shoots up to 0.8 m high should also be done in the first 1 to 2 years.

In Taiwan, quite detailed and severe pruning is used from a young age to assist tree size control and encourage regular flowering and cropping. This pruning is carried out after harvest and where a number of laterals arise from a main branch only one is let develop, and one from it and so on. The tree skirt is lifted quite high by removing all lower branches that bend down and touch the ground. The removal of a large amount of tip at harvest (up to 600 mm) is designed to encourage one strong flush to ensue, which will in turn carry the follow up crop. If pruning is not severe, several weak flushes result and flowering and cropping are poor. In addition if autumn flushes do occur after the one main flush required, these are pruned off and flowering will ensue, although it will be weaker and less prolific bearing.

Thinning out and opening up of the whole canopy is also done after harvest and quite large limbs may be removed in this exercise. Fertilizing at harvest or just prior, is done to encourage flush development after pruning.

In contrast, in India little or no pruning is done with the Indian selected varieties, and cropping is apparently heavier than in Taiwan where the plant is heavily manipulated.

With trees of 4 years old or older, girdling can be used to bring plants into cropping. This is done in autumn (March) in southern Queensland. In some areas, girdling has to be used yearly to ensure adequate flowering.

Root pruning in winter by cultivation to a depth of 10 cm is practised in China. Again this assists flower initiation. Cultivation to 5 cm is done at flowering and just prior to harvest when it accompanies fertilizing in China.

Irrigation

Irrigation is essential for lychee establishment and production. Withholding water in autumn and winter is desirable to discourage vegetative growth and to promote flowering.

Adequate moisture is essential particularly in the period from anthesis to 60 days after, to ensure adequate cell division in the lychee skin, to reduce splitting near maturity.

However, it is a personal opinion that some stressing is desirable during setting and fruit development to enable the plant to cope better with hot dry periods in spring and early summer in Queensland.

Fertilization

There appear to be as many fertilizer practices as there are lychee growers. We suggest during the first 3 to 4 years in the field the following programme for south Queensland conditions:

(1) Pre-plant: 0.5 kg single superphosphate applied to 1 m^2 at each planting site and dug in, together with 35 L of mill-mud or 10 L fowl manure (one month prior to planting).

(2) Post-plant: (Planting taking place in February to March).
1st year — 30 g urea per plant per month from the first flush after planting, plus 30 g of Fertica once each 3 months, with urea. Fertica has NPK analysis of 11.4:4.8:14.6% + S 7.6, Ca 4.3, Mg 1.3, B 0.1, Mn 0.1, Cu 0.04, Zn 0.02, Co 0.0006.
2nd year — In spring apply 25 L of mill-mud or 8 L of fowl manure. Increase urea to 40 g per plant and Fertica to 40 g per plant.
3rd year — In spring apply 35 L of mill-mud or 15 L of fowl manure. Keep at same level as for 2nd year if necessary, to ensure adequate plant size for cropping. Apply inorganics as for 2nd year.
In cool sites, do not fertilize in July and August, otherwise flushes will occur, which are more susceptible to frost damage.

(3) Three- or four-year-old trees of bearing age: Apply no fertilizer after spring of the 2nd or 3rd year depending on tree size, to ensure that slow down of vegetative growth will occur in the following autumn, to secure a good flowering. This is particularly the case where a lot of organics have been used previously, or on naturally fertile loam, clay loam or clay soils.

Water can be withheld over the same period and only enough water need be applied to prevent severe wilting.

Fertilizing bearing trees in Queensland is not known with any certainty. While the Chinese and the South Africans maintain fertilizing is required prior to flowering and prior to and just after harvest, the vigour of our trees suggest that we should not fertilize at flowering or in the spring period. We need to quieten down the spring flush which seems to compete with fruits that are trying to set and develop.

We suggest that fertilizer only be applied at or prior to harvest to boost the harvest flush, which emerges either just before or after harvest. If in doubt, do not apply any fertilizer if tree health and vigour are good. The lean and hungry lychee tree is the one that crops.

N, P, K, Ca, Mg and Zn deficiencies have been described in the lychee literature. South African leaf 'critical levels' developed as 'norms' from survey are N 1.3 to 1.4%, P 0.17 to 0.2%, K 0.8 to 1.2%, and the correct time to sample is April. Soil 'norms' quoted for South Africa are P, 5 to 8 ppm and K, 60 to 80 ppm.

Pests

Nematodes have been dealt with in the section on management. The following canopy pests are of importance in Queensland:

- Lychee erinose mite — *Eriophyes litchii**
- Soft scale — *Chloropulvinaria psidii* (Maskell)*
- Rhyparid beetle — *Rhyparida discopunctata* (Blackeb)
- Stem-girdler — *Trochilium chrysophanes* Meyr.
- Parasitic algae — *Cephaloures virescens**

Flower pests in Queensland:

- *Lycaenid* butterflies
- *Heliothis* sp.
- Orange fruitborer — *Isotenes miserana* (Walk.)
- *Tortricid* sp.
- *Lymantriia* sp.
- *Geometrid* sp.
- Fruitspotting bug — *Amblypelta nitida* Stal.*

Fruit pests in Queensland:

- Macadamia nutborer — *Crytophlebia ombrodelta* Low.*
- Lycaenid fruitborer — *Deudorix epijarbas diovis* Hew.
- Fruitsucking moths — *Othreis fallonia* Clerck. and *Othreis materna* Linneaus*
- Flying foxes (fruit bats)*

* These pests are more common and are generally more serious.
Control measures have been determined in all instances.

Diseases:

- Collar rot — *Botryodiplodia theobromae* (isolated occurrence).
- Post-harvest, *Glomerella cingulata, Phomopsis, Dothiorella* (occur in conjunction with insect damage).

Control measures have been determined in all instances.

Harvesting methods

Hand harvesting is done when fruits are full red ripe by clipping off the complete panicle carrying the fruits. For distant markets fruit may be picked before full colour is reached. Mechanical aids such as ladders or mobile picking platforms are required for older larger trees. Fruits wet from dew or rain should not be harvested. Fruits are clipped from the panicles in the packing shed leaving part of the fruit stalk attached to the fruit. Fruits are then sorted and damaged or stung fruits removed.

Marketing characteristics

The composition of the lychee fruit of the Kwai May variety is as follows: (Hamilton and Wenkam 1967).

Composition per 100 g of edible portion

Moisture 77.6%	Fibre 0.2 g	Vit. A. nil
Food energy (calories) 80	Ash 0.4 g	Thiamine 0.04 mg
Protein 0.9 g	Calcium 4.0 mg	Riboflavin 0.08 mg
Fat 0.3 g	Phosphorus 3.5 mg	Niacin 1.9 mg
Carbohydrate 20.8 g	Iron 0.4 mg	Ascorbic Acid 40 mg

Other similar tables of composition are noted elsewhere in the literature.

Fruits from different lychee varieties have variable recovery of flesh from seeds and skin. Fruits with small shrivelled seeds ('chicken tongued' seeds) as found in No Mai Chee and Kwai May are much preferred, as the edible portion of the fruit is much greater.

Fruits are marketed either in fresh state, at room temperature or pre-cooled to –1°C or as deep frozen fruit, dried fruit (Lychee nuts) or in cans.

In Queensland, fruits are sent off to markets in 5 kg cardboard cartons and more recently in strawberry punnets overwrapped with PVC clingwrap film.

Storage characteristics

The lychee quickly loses its bright red colour a few days after harvesting.

Various reports in the literature contain data on increasing storage and shelf life of lychee. Wax emulsion 9% plus 0.5% SOPP increases shelf life at 30 to 35°C and 59 to 60% RH by 3 days to 6 days.

Other reports indicate that at –1°C to 7.5°C lychees can be kept for 3 months thereby extending the lychee season considerably. However colour is lost in 9 to 10 days of storage. We have deep frozen lychees in polybags without any treatment and they keep well for up to 12 months at –15 to –18°C provided they are quick frozen in 2 hours or less.

South Africa commonly pre-cools to –1°C for shipping to the London market. The pre-cooling prevents fungal infection in transit. However, pre-cooled fruit should be kept in chilled show cases at retail outlets, and stored at home in domestic refrigerators until eaten.

The lychee is a non-climacteric fruit and thus partially ripe fruits do not continue to ripen once harvested.

Work in Queensland and northern New South Wales has shown that shelf life (with good skin colour) can be extended from 1 to 2 days to 8 to 9 days using a hot benomyl dip, air drying fruit and packing in plastic strawberry punnets completely overwrapped with PVC clingwrap film. Also, storage at 2 to 3°C will hold lychees in polythene film for 6 to 8 weeks quite easily and the fruit is still of good quality, although the colour deteriorates. (Brian Brown, personal communication).

Transport characteristics

Fresh lychees can be transported over long distances either by sea or air to distant markets, without serious depletion in quality.

However, care is needed in keeping out rotten and damaged fruits, and care in packaging and pre-cooling is also of considerable importance, to avoid fruit damage.

Packaging in polythene bags reduces darkening of lychee fruits in transit.

Irradiation of fruits with 25 krad required for disinfestation can be used on export lychees. These fruits should be packed in polythene and stored or shipped at 6°C. A shelf life of 4 weeks can be expected at 6°C from fruits treated in this way (Akamine and Goo 1977). Also fumigation with ethylene dibromide, as required by the U.S. Department of Agriculture, produces no detectable injury to quality or shelf life of lychee (Akamine and Goo 1977).

When lychees are wrapped in polythene or PVC film, care must be taken that the fruit is dry before packaging or problems will occur in storage or transit.

Processing characteristics

Lychees can be processed into a number of products, for example, canned lychees, lychee nectar, dried 'lychee nuts' and lychee wine as well as frozen lychees.

Chan and Cavaletto (1973) have developed a successful technique for lye peeling of lychee, which removes the skin prior to canning and eliminates the need for laborious hand peeling.

There are many references in the literature pertaining to drying of lychees to produce the dry lychee nut (for example Joubert 1970 and Kuhn 1962).

Dried lychees are often given as presents in China to gain face with their fellow man, and lychee wine was traditionally made for consumption at the 'Red Cloud Feasts'.

The prospects for developing processed products of lychee are good but while prices remain high for the fresh fruit, the economics for the exercise bears careful investigation. Canned lychees are quite inferior in flavour to fresh, frozen lychees or dried lychees.

References

Akamine, E.K. and Goo, T.T.S. (1977), Effects of gamma irradiation on shelf life of fresh lychees *(litchi chinensis* Sonn.*)*. Hawaii Agric. Exp. Stn. Res. Bull. 169.

Chadha, K.L. (1968), Litchi cultivation in India. *Indian Hort.* **12,** 13.

Chan, H.T. and Cavaletto, Catherine G. (1973), Lye peeling of lychee. Hawaii Agric. Exp. Stn. Res. Bull 215.

Cull, B.W. and Hams, F.D. (1974), Litchi finds a new home in Queensland. *Queensland Agricultural Journal* **100,** 597.

Goto, Y.B. (1960), The lychee (litchi) and its processing. Pacific Rim Food Conference 1, p. 15.

Groff, G.W. (1921), 'The lychee and longan'. Orange Judd Co., New York.

Groff, G.W. (1944), Some ecological factors involved in successful lychee culture. *Proceedings of the Florida State Horticultural Society* **56,** 134.

Hamilton, R.A. and Wenkam, Nao S. (1967), Nutritive values of major fruit crops grown in Hawaii. *Proceedings XVII International Horticultural Congress* **4,** 107.

Hsu, H.H.; Peng, C.; Hsuch, J.H. and Tsai, Y.F. (1964), The discovery of wild lychee trees. *Acta Hort. Sinica* **3,** 415 (Chinese).

Joubert, A.J. (1970), The litchi. Citrus and Sub-Tropical Fruit Research Institute Bull. 389.

Kuhn, G.D. (1962), Dehydration studies of lychee fruit. *Proceedings of the Florida State Horticultural Society* **75,** 273.

Nakata, S. and Watanabe, Y. (1966), Effects of photoperiod and night temperature on the flowering of *Litchi chinensis. Botanical Gazette* **127,** 146.

Singh, L.B. and Singh, U.P. (1954), 'The litchi *(Litchi chinensis* Sonn.*)*'. Superintendent of Printing and Stationery, Lucknow U.P., India.

Storey, W.B. (1973), The lychee. California Avocado Society Yearbook 1972–73, p. 75.

Storey, W.B.; Hamilton, R.A. and Nakasone, H.Y. (1953), Groff — A new variety of lychee. Hawaii Agric. Exp. Stn. Circ. 39.

Further reading

Anon. (1976), Bibliography of litchi (lychee) 1916–1976. Qld. Department of Primary Industries, Central Library.

Anon. (1980), Litchi: Annotated Bibliography (1974–79). *Qld. Department of Primary Industries Bibliography Series* 80–28.

LONGAN by B.J. WATSON
Euphoria longan (Lour.) Steud.
Nephelium longan Lour.

Botany

The longan was earlier classified in the genus *Nephelium* as *Nephelium longana* Camb. but is now generally accepted as *Euphoria longan* (Lour.) Steud.

The basic morphology of the longan is similar to other sapindaceous trees such as lychee, rambutan, pulasan etc., but most particularly related to the first.

Groff (1921) relates the description by a Singapore Chinese gentleman referring to the differences between fruit of the four most popular sapindaceous species.

'The rambutan is the long haired gentleman of the group, the pulasan having a modern cut, the litchi closely cropped and the longan as bald as the venerable Buddist monk'.

The genus *Euphoria* consists of 7 species, but only *E. longan* is cultivated to any extent. *E. didyma* (the alupag) is an interesting but non-commercial species native to the Philippines.

Longan was formerly reported to be indigenous to the lowlands of Ceylon, southern India, Burma and China, (Chandler 1958) but the area of distribution is now thought to include the medium elevations (up to 1000 m) of the mountain chain from Burma to southern China (Pholprasit, personal communication 1979). Principal areas of cultivation are in Thailand, China and Taiwan. Hong Kong, Burma, India, Laos, Vietnam and Florida also have minor industries. Common names are lungan (dragon eye) and longyen (China-Taiwan), lamyai (Thailand) and lengkeng (Malaysia-Indonesia). Southern China is thought to have been the centre of earliest selection of clonal material and written descriptions date back as far as 1000 A.D.

The tree is of medium size, 10 to 20 m high tending to a hemispherical crown, although more upright when grown from seed. Bark is corky instead of smooth as in the case of lychee, and the wood more brittle. Leaves are petioled, alternate and paripinnate with 6 to 9 pairs of leaflets arranged opposite and alternate on the rachis. Leaflets are glossy dark green on the upper surface and paler on the lower, commonly up to 120 mm long. Young leaf flushes are often red. Inflorescences are terminal, leafless, erect and with widely branched panicles up to 300 mm long. Flowers are small and yellowish brown, with five, sometimes six, petals. Filaments are pubescent and anthers glabrous. The ovary is divided into two, sometimes three parts (Groff 1921).

Lian and Chen (1965) describe longan flowering in each panicle as a progression of opening of staminate (pistil non-functional) then pistillate (stamens non-functional) followed by hermaphrodite and finally, again, staminate flowers. As with lychee there is overlap of flower types on one tree since not all panicles develop at the same time. Fruits take from 5 to 7 months to mature according to cultivar type and climatic conditions. The fruit is classified as a nutlet, globose (but sometimes with pronounced shoulders), dull green while developing and ripening to a dull yellow-brown. The panicle may carry up to 80 individual fruits which vary in weight from 5 to 20 g, but with premium commercial grades in the 12 to 18 g range (Vangnai personal communication 1979). The pericarp is thin but tough and leathery. The aril is white to off-white and may be up to 75% of the total fruit weight. The seed is small, round and black and comes substantially free from the aril. Total soluble solids (sugar) content of the aril ranges from 15 to 25% (Knight *et al.* 1968).

Climatic requirements

The prime environment for commercial longan production lies between latitudes 15 to 28° north and south of the equator, but variations in altitude and continental climatic influences obviate a rule of thumb definition.

Longans are most successful in areas with short, but cool winters (temperature normals 15 to 22°C for 2 to 3 months) followed by high temperatures in spring and summer. The limitation within these zones is the incidence of frost. Campbell *et al.* (1977) rated frost damage in Florida as slightly more severe on longan than lychee, with young trees being killed at –1°C. Flowers were damaged at 0°C. There was much less permanent damage to longans than to lychees in Hong Kong following a –3°C frost although leaf damage was just as severe (Anon. 1963). Minimum temperatures required to induce flower differentiation would appear to be higher than that for most lychee varieties, but a satisfactory definition of chilling requirements is not available. In Thailand, commercial longan production is in warmer (lower elevation) areas than those of the best lychee farms. However, in China, Groff (1921) describes natural distribution of wild types and cultivation at higher elevations and latitudes than the lychee in Kwangtung, Fukien and Szechuan provinces.

Longans are seldom irrigated in Asia, and most production areas have annual precipitation in the range of 1000 to 2000 mm. However, at the lower end of this range best production is on alluvial soils with a relatively high water-table. High soil moisture is essential for a satisfactory fruit set and filling.

In Thailand, China and Taiwan there is significant rainfall (greater than 100 mm per month) during the peak fruit filling months of May to September. In north Queensland substantial rain may not occur until a much later stage of development so that on most soils supplementary irrigation is a necessity.

Soil requirements

Deep rich moderately well-drained alluvial clay loam soils are preferred. In China and Taiwan, shortage of arable land has forced longan production onto hillsides but their yields are not as substantial as on alluvial soils. Groff, however, states that in China trees grow better on higher, freer draining soils than on alluvial soils. This statement may be due to the fact that lychee has taken priority for the better soils.

Genetic diversity

Longan has been selected and asexually propagated in China for some hundreds of years. Selection in Thailand, China, Taiwan and Florida has tended towards large fruit, high percentage aril recovery, good flavour, crisp flesh (particularly in Asia), high sugar content, regular and heavy yields, a long seasonal cultivar distribution of harvest and good shelf life. More recently processing requirements have influenced selection for pure white aril. Biennial bearing has been the main obstacle in selection. Seedling progeny are extremely variable and small fruit size appears to be a dominant gene influence. However, some of the heaviest and most regular bearing trees are small fruited types. Knight *et al.* (1968) also indicates that good flavour is most commonly but not necessarily associated with high sugar content.

The best Thai clones have different names from those in China, although it is possible that many are identical. Thai production commenced only in the late 1800s — originating from plants brought in from China (Boonrawd, personal communication 1979).

The most popular longan cultivars in various countries are:

China	—	Shin Hsia, Wu Yuan, She'P'i, Kuan Yen, Ts'ao P'u Chung, Niu K'ou Tze, Pa Yueh Hsien, Wu Yuan No. 106, Hua Kou, Fu Ling, Huang Kou, Fu Yen, Wu Lung Ling, Yu T'an Pen, P'u Ming An, Ch'ih K'o, Tung Pi, Chi Tan and Chiu Yueh Wu.
Taiwan	—	Aliou, Carambola Leaf, Duan-Yu, Chien Liou, Fu Yen, Tiang Suan Jou, Nu Tze Yen, Tai Fun Ke, Tun Shan Fun Ke, Tai Ching Ko, Si Kung Mou, Yeung Tau Yeh, Yick Liao, Shao An Liao, Hu Te, Ko Ping, Lung Muk and Ching Shan.
Hong Kong	—	Shek Yip, Fa Haak and Fa Hok Chai.
Thailand	—	Daw (E-Daw-Kiew and E-Daw-Dang), Chompoo (Seechompoo), Haew (E-Haew-Yod-Kiew and E-Haew-Yod-Dang), Biew Kiew (Biew Kiew Kan-oon and Biew Kiew Kan-Khang), Dang (E-Dang Hua-Boom and E-Dang Klom), Baidam, (Baidum), Luang and Talub Nak.
Florida	—	Kohala, Chompoo, K. Sweeney, Blackball, Ponyai, Kona, Homestead No. 1, Homestead No. 2 and Dagelmen.
Hawaii	—	Fukho No. 2, Kohala, Ilao, Wai, Carambo.

Production season and time of cropping

Flowering of longans in north Queensland normally commences in late August to early September, continuing to October according to the cultivar and seedling type range. Harvest is from mid-January to mid-March with the peak in February. In equivalent months, flowering is earlier in Thailand and China (18° to 28°N), occurring in January to March and the harvest range is from late June to late September. The period of fruit development in the latter areas varies from $5^1/_2$ to 7 months.

Yield potential

In Thailand the average yield per hectare per annum is 6.0 tonnes (120 kg per tree) for trees of all age groups included. However, there is considerable variation from year to year and biennial bearing trends coincide for most varieties (Vangnai, personal communication 1979). For 12 to 15-year-old trees the yield range in any year is from 6 to 170 kg per tree, there being an average of one poor, one fair and one excellent crop in three years. Individual trees may live to an old age and (although at lower density) have recorded up to 1400 kg in one season. Small fruited inferior types are reputed to bear more heavily (300 to 400 kg) and consistently. Yields from selected varieties in Taiwan are reputed to far exceed Thai averages; however, densities (100 to 120 trees per hectare) are greater than the Thai average of 50 trees per hectare.

Air layered trees normally commence bearing in the third or fourth year after field planting in north Queensland.

Market acceptability

The longan flavour is musky and distinctive, quite unlike lychee and rambutan. For most Europeans, the fruit usually develops appeal after a number of tastings, somewhat like the acquired taste many people develop for fruit such as avocado after a period of time. Few of the seedlings growing in Queensland have the aril thickness, texture and flavour of the best Thai and Chinese varieties and thus can create an unfavourable impression. Thai production (in excess of 150 000 tonnes per annum, about one quarter of which is exported) attests to the very strong Asian acceptance. Unfortunately Queensland production is in summer, in competition with a wide range of deciduous fruits. However, the demand for longan from the Asian population in Australia far exceeds supply, and even inferior seedling types are accepted.

Export potential

There is virtually no production of longan in the southern hemisphere and the Asian market must be considered a very promising area. Out of season fruit would undoubtedly realize very good prices, as is the present situation with Australian mangoes. Purchase by only 1% of the south-east Asian population would provide a substantial demand.

Cultural details

Propagation

Nearly all Thailand's longan trees are propagated as air layers during the wet season. However, Boonrawd (personal communication 1979) suggests that grafting on to longan seedlings does give greater strength to the tree. Groff (1921) stated that in Kwangtung (Guangdong) province in China most trees were inarched (approach grafted). The seedling stock trees were grafted in the field at about 3 to 5 years old. Seedlings from the cultivar Wa Yuan (blackball) were most commonly used for rootstock. Further, the Chinese farmers believed that double or triple grafting (successive grafting operations on top of the previous graft) gave added strength to the tree and produced larger and sweeter fruit. In north Queensland air layering with the peat and polythene wrap method is effective, although not as successful as with lychee. Best results are from air layering in March and April. Air layered trees at Kamerunga often blow over in cyclonic winds when 5 years old or older. Some recent cultivar accessions from Thailand and Florida have been approach or wedge grafted, but are not old enough to determine stability in high winds. Scion material for wedge or whip grafts is preferably taken from semi-mature wood back from the branch tip and particularly from the area of close internodes. Cincturing a month prior to grafting is recommended in Florida (Knight 1969). The modified Forkert bud graft is also a very acceptable technique.

Cuttings in mist and bottom heat are moderately successful, but I.B.A. and N.A.A. are not beneficial (Anon. 1978). Hard, thick, terminal cuttings about 250 mm long with full leaf cover taken in June to August have provided best results at Kamerunga Horticultural Research Station.

Management

Thai growers remove all watershoots and surplus branches within the tree canopy, and as well, maintain the skirt at least 1 m above the ground. After harvest, any panicle remains are pruned off, but it is suggested that cutting back too hard leads to a poor crop the following year. The problem with biennial bearing (which is common to all but a few cultivars) is sometimes lessened by removing some panicles after fruit set in the 'on' year. This also tends to increase fruit size in the balance of the crop. Biennial bearing, however, is largely inevitable since the crop matures in mid to late summer and there is only a short period of suitable temperature to provide new growth before winter. The new growth carries the following year's crop. Fruit set and retention is not such a problem with longan as with lychee and Groff (1921) reported that it was common practise in Fukien province to remove three-quarters of the flower panicles. This obviously reduces biennial bearing.

Thai trees are susceptible to strong wind damage and are frequently permanently propped or soil is progressively built up around the butt and lower branches to increase stability. Curiously this practice does not lead to trunk death or root rots providing the soil mound (up to 2 m high and 3 m diameter) is built up slowly over a period of years.

Irrigation

Very little information is available. As stated previously, the wet monsoon season in Thailand, southern China and Taiwan commences just before, at, or shortly after flowering and except on the lightest soils in dry zones there is no soil water deficit during fruit filling. However, in Queensland supplementary irrigation is required from at least September to December.

Fertilization

Very little written information is available. Thai farmers have low inputs and use mainly wood ash and peanut husks. Some liquid proprietary fertilizers are applied as a soil drench in April and December.

In southern China Groff reported application of night soil at flowering, after harvest (August), and in December. Inorganic fertilizers (mainly N:P:K — 12:12:17) are used at similar timings in north Queensland.

Pests and diseases

Insects pests of note in Hong Kong, China, Laos and Thailand are:
hemiptera —*Pyrops candelaria; Tesseratoma javanica;* and *T. papillosa* (stink bug);
lepidoptera — *Setora nitens; Clania* spp; *Archips micacaena;* and *Olethreutes discana;*
coleoptera — *Aspidomorpha sanctaecrucio* (Dean 1978).

In Thailand there are problems with flower caterpillars, fruit fly and fruit bats. Fruit panicles are sometimes bagged to prevent fruit fly attack. In addition, the 'witches broom' mycoplasma causes growth malformation and fruit drop (Smitamana, personal communication 1979).

In north Queensland the longan is still substantially free of pests and diseases as compared with lychee, although the macadamia nutborer *(Cryptophlebia ombrodelta)* can cause significant fruit damage.

Harvesting methods

Maturity is gauged by fruit shape, skin colour and taste. In Asia, part or whole panicles are cut or broken from the tree. However fruit are not allowed to fall to the ground.

Panicles are sorted and subpanicle groups of fruit detached leaving 100 to 200 mm of stalks attached. Stalked fruit are reputed to remain fresher than separated fruit after harvest. Thai growers pack fruit in 23 to 24 kg woven baskets.

Marketing characteristics

Thailand has established fruit grades, grade A for 55 to 75 fruit per kg (18 to 13.3 g each), and grade B, 76 to 80 fruit per kg (13.2 to 12.5 g each). A third grade encompasses most small-fruited seedling types.

Although fruit are mostly marketed with stems, with increasing use of refrigeration and greater consumer sophistication there is less resistance to purchasing stemless fruit. Packaging in standard Australian type cartons is, however, still unheard of.

A summary of fruit characteristics encompassing both unselected seedlings and proven clones is shown in the following table.

Thai longan fruit characteristics

Fruit per panicle	35 to 80	T.S.S. 15 to 25° brix
Mean total fruit weight	4 to 20 g	Fruit diameter 10 to 35 mm
Mean aril recovery	40 to 75%	Pericarp colour, yellow to light brown.

(Boonrawd, personal communication 1979)

Storage characteristics

There has been very little work in this area. In general, the fruit store longer under refrigeration than do lychees and there is no significant colour change in the pericarp nor a substantial tendency to post-harvest rots. In South East Asia fruit are kept on ice for up to 1 week without major loss of market appeal.

The fruit stores well if quick frozen and there is no contamination in flavour from the pericarp. Aril texture and colour after thawing are quite acceptable.

Transport characteristics

The fruit has a tough pericarp which withstands bruising. There are no evident transport problems.

Processing characteristics

There are substantial canning industries in Thailand, China and Taiwan. Large fruit are used, preferably those with small seeds. Due to high brix levels little sugar additive is required and fruit can be canned in their own juice. Canned longans retain their individual flavour better than do rambutan and lychee.

Historically China's production was predominantly processed as dried fruit, either intact or with the pericarp removed. However, except for medicinal use fruit are now mainly eaten fresh or are canned.

References

Anon. (1963), Frost susceptibility of litchi, longan and guava. Hong Kong Department of Agriculture and Fisheries Annual Report 1962-63, p. 31.

Anon. (1978), Litchi and longan propagation. Hong Kong Department of Agriculture and Fisheries Annual Report 1977-78, p. 13.

Campbell, C.W.; Knight, R.J. and Zareski, N.L. (1977), Freeze damage to tropical fruits in southern Florida in 1977. *Proceedings of the Florida State Horticultural Society* **90,** 254.

Chandler, W.H. (1958), 'Evergreen orchards'. Lea and Febiger, Philadelphia.

Dean, G.J. (1978), Insects found on economic plants other than rice in Laos. *Pans* **24,** 129.

Groff, G.W. (1921), 'The lychee and lungan. Part 1'. Orange Judd Co., New York.

Knight, R.J., Manis, W.E., Kosel, G.W. and White, C.A. (1968), Evaluations of longan and lychee introductions. *Proceedings of the Florida State Horticultural Society* **81,** 314.

Knight, R.J. (1969), The longan, a subtropical fruit with potential in Florida. Rare Fruits Council, South Fla. Inc. Fact Sheet.

Lian, T.K. and Chen, L.L. (1965), A preliminary observation of flower bud initiation in the longan in Foochow. *Acta Hort. Sinica.* **4,** 12.

RAMBUTAN
Nephelium lappaceum L.

by B.J. WATSON

PULASAN
Nephelium mutabile Blume

Botany

Rambutan is native to west Malaysia and Sumatra. Closely related species are found throughout east and west Malaysia, Borneo, Sumatra, Java, Philippines, Thailand, Kampuchea, Vietnam and Burma. Common names: rambutan (Malaysia, Indonesia, Philippines and English speaking countries); rambutao (Spanish); litchi cheval (French); and ngo phruan (Thailand).

Pulasan has origins in west Malaysia, Borneo and Sumatra. Common names: pulasan (Malaysia and English speaking countries); kapulasan (Indonesia); and ngo khonsan (Thailand).

RAMBUTAN
Under normal conditions it is a tree of medium size from 12 to 25 m high and canopy width approximately two-thirds the height. However, there are variations in form in the cultivars — from erect to lax weeping crown. Seedlings are mainly erect with a straight trunk and a dense head of branches. Leaves are petioled, alternate and pinnate with two to four pairs of leaflets arranged alternately or sub-opposite on the rachis. Inflorescences are both axillary and terminal, erect and widely branched. Flowers are small, greenish pubescent, 600 to 2000 per panicle, and usually functionally unisexual.

Valmayor *et al.* (1970) classify trees in three groups according to flower characteristics.

(a) Those producing only staminate flowers — male trees (usually 40 to 60% of any seedling population).
(b) Those producing hermaphrodite flowers which are functionally female.
(c) Those producing hermaphrodite flowers, some of which are functionally female and some which are functionally male.

In type (c) which is the most common form in cultivar selections, the percentage of male functional flowers ranges from 0.05 to 0.9.

The fruit is classified as a nutlet, globose to ovoid, green in colour while developing and ripening to various shades of pink, red or yellow. Size (excluding spines) ranges from 30 to 80 mm long by 25 to 50 mm wide. The pericarp is thick and covered with closely spaced tubercles, each terminated by a soft pliable spine of variable length. The edible aril is translucent white to yellow-white. The aril thickness (8 to 15 mm) and eating quality vary according to the cultivar. The seed is elliptic ovoid or oblong 20 to 35 mm long by 12 to 22 mm wide. The aril may or may not peel cleanly from the seed. In 'freestone' cultivars the aril usually detaches together with a somewhat objectionable seed testa.

Fruit weight is 20 to 60 g, of which pulp (aril) weight is 30 to 58%, pericarp 40 to 60%, and seed 4 to 9%. Most cultivars have high ascorbic acid content.

PULASAN
Tree form and foliage are similar to rambutan but the main distinguishing characteristics are: tree is smaller, 10 to 15 m high, leaflets narrower, form (branches) more open and less fruit on the panicle. In the fruit, the pericarp is thicker, usually dull red with spines or tubercles very much shorter. The aril is usually white, clinging to the seed and 35 to 45% of total fruit weight. Fruit size range is 40 to 70 mm long

and 40 to 60 mm in diameter. Flavour and quality of the aril is invariably good and often preferred to that of the rambutan.

Climatic requirements

Both species are currently grown commercially within 15° of the equator but most commonly within 12°. According to Terra (1952) they can be grown well at altitudes of up to 500 m within 8° of the equator. The mean yearly temperature normals within Asian areas of production are 22 to 30°C.

In Queensland (Cairns, coastal area) trees are actively vegetative in months with temperature normals above 22°C. However, one cultivar (R134) is vegetative year round. At Nambour (26°S) growth is restricted to the summer and autumn months and defoliation is common in winter.

The crop is grown in high rainfall areas in Asia — 2000 to 5000 mm per annum. Supplementary irrigation from flower set to harvest is provided at Chanthaburi in southern Thailand, in a 3000 to 4000 mm precipitation zone. Irrigation would be necessary in all potential production areas in Australia.

Leaf tip die-back and marginal necrosis during winter in Queensland is thought to be due to a combination of low humidity and low temperature. Pulasan is less tolerant of high levels of radiation, particularly during field establishment. Both species are intolerant of exposure to wind, particularly under conditions of low relative humidity. Best growth in Queensland is in protected situations.

Soil requirements

Deep, rich, well-drained sandy loams or clay loams, with high organic matter content are generally preferred. Alluviums or residual soils derived from basaltic rocks are most commonly used. A pH range of 4.5 to 6.5 is preferred, since at higher levels Fe and Zn deficiencies are common. At Kamerunga organic mulches for the first 4 years after planting have proven very beneficial.

Genetic diversity

In rambutan, variation exists in tree form and growth; leaf colour and size; flower type and number per panicle; fruit pericarp and tubercle colour, thickness and length; aril colour, thickness, texture, brix and acidity; adherence of aril to testa; seed size and form; susceptibility to pests and diseases, and drought tolerance.

The main clonal selection has been carried out in Malaysia and Indonesia and to a lesser extent in Thailand and the Philippines. Malaysia, with the greatest diversity of types has restricted commercial recommendation to 15 clones. Pulasan varies much less than rambutan in all respects. Some 38 clones from the various countries have been imported into Queensland.

The most popular rambutan cultivars in various countries are:

Malaysia — R3 (Peng Th'ng Bee), R4 (Ya Tow), R7 (Kepala Besar), R9 (Tau Po Cheng), R99 (Triang), R134, R137, R139, R156, R160 (Khaw Tow Bak), R161, R162 (Oh Heok), R163 (P.J.), R165, R167, R168 (Che Emhong), R170. (Deli Cheng) and Hew. Current commercial recommendations for fresh fruit cultivars are R3, R134, R156, R161 and R170.

Thailand — Chompoo (Seechompoo), Rongrien, Bang Yi Khan, See Tong and Nam Tan Kruad.

Indonesia —	Atjeh Kering, Manis, Lebakbulus, Silengkeng, Pao Pao, Sinyonya (Seenjonja), Aceh Garing, Aceh 6B, S.K.W.L., Aceh Padang, Biji, Aceh Gundul, Aceh Kuning, Sitangkue, Simacan (Seematjaan), Binjai, Aceh Gendut, Aceh Rapiah, Aceh Gedong, Padang Bulan and Gula Batu.
Singapore —	Jit Lee (Deli), R37, Ah Kow Cheng and Kim San Cheng.
Philippines —	Seematjaan, seenjonja, Maharlika and Maharlika 'Belen'.

Most widely planted pulasan cultivars are:

Malaysia —	P1, P4, P5, P6, P8, P22, P28, P54, P63 and Lee.
Indonesia —	Merah, Poetih, Sibabad and Koeneng.
Philippines —	Sibabat (Sibabad).

Production season and time of cropping

There appears to be no photoperiod or temperature requirement for flower initiation. Differentiation is reputed to precede flower panicle emergence by only 3 to 4 weeks. In tropical countries flowering usually occurs during or at the end of a 'relatively' dry period. Commonly there are two crops per year but in most cases from different terminals on the tree. In southern Thailand (12°S) trees are managed (pruning, fertilizing, watering and flowering panicle removal) in order to provide a single major crop. However, there is cultivar as well as climatic seasonality in production.

The period from anthesis to harvest is normally 100 to 120 days but following autumn flowering in north Queensland this period may extend to 160 days. In Cairns, flowering peaks occur in July to August and December to January, but may be as late as April. Thus there may be fruit at any time between December and August.

Yield potential

Rambutan production for the range of cultivars in north Queensland has yet to be determined, although at Kamerunga cultivar R168 has yielded 88 kg in year 6. Table 2 denotes production from (i) Chanthaburi, Thailand (Prabhadhapongse, personal communication 1979); and (ii) Serdang, Malaysia (Loh and Lee 1966).

Planting densities are commonly 80 to 120 trees per hectare. Pulasan yields are generally much less than rambutan.

Market acceptability

Rambutan and pulasan are attractive eye catching fruit and the slight 'bite' of acidity in the aril is preferred by some to the extreme sweetness of lychee. Asians have a penchant toward crisp fresh freestone varieties irrespective of presence of any objectionable testa. Europeans prefer sweet stringless fruit without objectionable testa and if such cultivars are produced in Australia then very good market acceptance is assured. The cultivars R3 and Rapiah are relatively testa free.

Pulasan generally have less problems with testa than rambutans. Production outside periods of peak supply of other fruits is desirable.

Table 1. Characteristics — Rambutan cultivars at Kamerunga Horticultural Research Station

Cultivar	Tree size	Fruit pericarp colour	Mean fruit wt (g)	Mean % aril by wt	Mean % pericarp by wt	Mean % seed by wt	Brix *
R3	medium	red	36	41	52	7	22.5
R4	large	red/crimson	26	36	58	6	18.6
R156	medium	yellow/orange	35	56	39	5	19.9
R168	large	crimson	35	37	56	7	20.5
Chompoo	small	orange	42	44	50	6	19.0
'Winkel'	large	red	48	33	61	6	20.0
'Stephens I'	large	crimson	45	40	53	7	21.0

* N.B. Fruit not harvested in same month and thus variation may be due in part to time of year.

Table 2. Yield fresh fruit, kg per tree

	Years after field planting						
	3	6	9	12	15	18	21
(i)	20	100	200	300			400+
(ii)	15	30	55	85	165	250	300+

Export potential

There is a relatively low level of export from rambutan producing countries to Europe and Japan as compared with mango and some other fruits. This may be due to lack of cultivar selection and production for these market tastes as well as to the short post-harvest life.

Australian export potential cannot be gauged at the present time.

Cultural details

Propagation

Cultivar propagation in Asia and Australia is mainly by the modified Forkert bud graft and approach grafting onto vigorous seedlings. Air layering is limited due to high mortality. Use of cuttings has not been thoroughly researched but pre-cincturing appears necessary. The best tree shape (wide crotch angles) is obtained from the Forkert bud graft. In north Queensland bud grafting has best results from October to December and April to May but for the latter period is complicated by low temperatures at the stage of bud forcing unless bottom heat is provided. However, forcing of buds can be delayed until the spring.

Management

Pruning out of all water shoots and dead branches within the tree frame is common. Panicle remains are cut off after harvest to induce strong canopy regrowth. Inter-tree ground cover is usually grass or legumes. Wind-breaks are essential in north Queensland. For cultivars with a very low or negligible percentage of male functional flowers a spray with sodium N.A.A. at 10 to 50 ppm on panicles a metre apart on the tree at the time of early flowering converts sufficient flowers to male functional status for adequate pollination.

Irrigation
Prolonged water stress at any time is detrimental and irrigation may be essential in the period from flowering to harvest. Soil moisture deficit gives a poor flower set and 'flat' (poor aril development) fruit. A suggested watering regime is to 85% of pan evaporation.

Fertilization
Nutrient removal by the crop is relatively low. Total element removal is shown in table 3.

Table 3. Nutrient removal (kg per ha) from harvest of 7300 kg of fruit and panicle twigs per hectare (Ng and Thamboo 1967)

N	P	K	Ca	Mg
15	2	11.7	5.9	2.7

Malaysian fertilizer recommendations (Kanapathy 1976) for poor soils are N:P:K mixtures approximating 14:4:3.5 up to first fruiting and thereafter 12:4:7 in the progressive increase range of 0.1 to 4 kg per tree per annum from planting up to 20 years of age.

In Thailand (Prabhadhapongse, personal communication 1979) fertilizers are applied to the drip line three times a year and chipped in; as flowers emerge N:P:K — 1:1:1 or 1:1:2; just after harvest 2:1:2 and chicken manure; after pruning 1:1:1 or 2:1:2. Low rates of dolomite application are recommended in Malaysia, but both lime and dolomite are regarded as detrimental to rambutans in Thailand, particularly for young trees.

Pests and diseases
Diseases in Asia which are of a serious nature but not present in Australia are *Dolabra nepheliae* (stem canker) *Thyronectria pseudotrichia* (stem die-back) and *Oidium nephelii* (flower, fruit and leaf mildew). The most serious pathogen in north Queensland is *Botryodiplodia theobromae,* a trunk canker originating in the graft union area, which can kill trees if left untreated.

Fruit flies, including the Oriental fruit fly are not troublesome in Asia unless overripe fruit are left on the tree. In Queensland, beetles of *Rhyparida* spp., larvae of various lepidoptera damaging leaves and flowers, and mealy bug infestation of fruit present the most problems. Fruit bat damage is sometimes serious and difficult to control.

Harvesting methods

For each variety, maturity is gauged by development of pericarp colour and aril taste. Individual fruit or whole panicles are cut off and in Malaysia and Indonesia presented for sale in bunches of panicles. In the Philippines and Thailand individual fruit are detached prior to sale. On large trees hooked knives on poles are used to detach the panicles. The arrangement of tubercles and the thick pericarp of fruit prevents major damage when panicles fall to the ground although any damage is undesirable. In order to harvest all fruit at correct maturity it may be necessary to pick trees at 3 to 4-day intervals over a period of 2 to 6 weeks.

Marketing characteristics

Fresh fruit are very attractive and eye catching particularly those with pink or red pericarps. Under ambient temperatures (25 to 33°C) and exposed to the air, fruit remain attractive for only 1 or 2 days and usually stay marketable for only 3 to 4 days. Removal of the panicle stems hastens deterioration.

Storage characteristics

Fruits are non-climacteric, showing a high respiratory activity after harvest but this declines with storage period. Mendoza *et al.* (1971) states that ripe fruits at 25 to 30°C may remain marketable for 6 days if stored in sealed or perforated polythene bags. The best storage temperature, however, is 10°C at which the fruit may remain marketable for in excess of 12 days in sealed polythene bags or up to 10 days in perforated bags. Relative humidity during storage should be 75 to 95% and addition of benomyl (1000 ppm a.c.) as a pre-pack dip is very effective in preventing storage rots even at temperatures of 26 to 30°C. Controlled atmosphere may assist with preventing weight loss in storage.

Transport characteristics

Fruit handle extremely well during transport if packages are reasonably rigid.

Processing characteristics

There is a substantial canning industry in Malaysia and Thailand. Brix readings for ripe fruit range from 14 to 20° but sugar solution is normally added. Thai canners buy fruit on the basis of minimum size qualification and not cultivar or aril peeling characteristics. There is no information on frozen products. At the home level fruits are used for jams and jellies.

Components of fresh fruit	per 100 g of aril
Water	83 g
Calorific value	63 cals
Protein (N x 6.25)	0.8 g
Total Carbohydrate	14.5 g
Fat	Trace
Calcium	25.0 mg
Iron	3 mg
Vitamin C	20 to 45 mg

References

Kanapathy, K. (1976), Guide to fertilizer use in peninsular Malaysia. Bull. 143, Ministry of Agriculture and Rural Development, Malaysia, p. 88.

Loh, C.L. and Lee, C.S. (1966), Rambutan yields. Malaysian Ministry of Agriculture and Co-op. Divis. Agric. Dept. Conf. Information Paper 52.

Mendoza, J.R. (Jr), Pantastico, E.R.B. and Javier, F.B. (1971), Storage and handling of rambutan. *Philippine Agriculturist* **55,** 322.

Ng, S.K. and Thamboo, S. (1967), Nutrient removal studies on Malayan fruits: durian and rambutan. *Malaysian Agricultural Journal* **46,** 164.

Terra, G.J.A. (1952), Some ecological requirements of Indonesian fruit trees. *Landbouw* **24,** 193.

Valmayor, R.V., Mendoza, D.B. (Jr), Aycardo, H.B. and Palencia, C.O. (1970), Growth and flowering habits, floral biology and yield of rambutan. *Philippine Agriculturist* **54,** 359.

TAUN or DAWA

by B.J. WATSON

Pometia pinnata J.R. & G. Forst.

Botany

In the family Sapindaceae, the genus *Pometia* includes some 10 tropical tree species originating in the region from Ceylon and the Andamans throughout South East Asia including Papua New Guinea and the Pacific to Fiji and Samoa. However, it is generally now accepted that there is only a single species, reduced to several forms (Anon. 1977). The best known forms are var. *pinnata,* var. *acuminata,* var. *alnifolia* and var. *glabra.*

Pometia pinnata is most commonly found in Papua New Guinea, the Solomon Islands, Western Samoa and Fiji and the fruit used most significantly in Papua New Guinea and Fiji where it is known as taun and dawa respectively. Other 'local' names are kasai, nalugai, truong, tava and matua. It is also often referred to as Fiji longan but this name is not appropriate.

As a forest tree *P. pinnata* is most commonly known for its timber, and trees vary in height from 25 to 40 m with a trunk diameter of up to 900 mm. Trees tend to be erect even when planted in the open. Leaves are large (up to 900 mm long) and pinnate with leaflets 300 mm long by 75 mm wide. Leaflets are tomentose, light green and new flushes are deep wine red. As with some sapindaceous species, flower type is variable. Flowers are effectively unisexual. There may be trees entirely functional male, and others with predominantly functional female, but including some functional male flowers on the same panicle (Jacobs 1962). Panicles develop to 500 mm, with small cream-white flowers. Newly set fruit may be vivid red, then change to green and then pale red when ripe (Whitman 1972). Other forms are light purple or dull yellow when ripe (Jacobs 1962). Borne in clusters on the panicle, individual fruit have a smooth pericarp and are round to oblong and up to 70 mm in diameter. The aril is semi-transparent, white and quite juicy. The seed is round, up to 25 mm in diameter. Fruits weigh from 45 to 80 g each. The pericarp is hard and approximately 4 mm thick.

Climatic requirements

The tree has a natural distribution in tropical areas within latitudes 20° north and south of the equator. It is mainly confined to zones with a minimum of 2500 mm mean annual precipitation fairly evenly distributed.

Temperatures in the Cairns area of north Queensland would appear to be quite satisfactory for growth. Two seedling trees at Kamerunga (one each from Papua New Guinea and Fiji) have grown 6 m in 2 years. In contrast, Whitman (1972) states that trees in southern Florida (26°N) grow approximately 500 mm per year but are usually retarded by frost damage. Within 12° of the equator the tree occurs naturally up to 1200 m altitude and thus must be fairly hardy.

Soil requirements

P. pinnata is most common in poorly drained sites (Anon. 1977) in river flats, but occurs as well on open ridges and spurs in lowland rainforests.

Genetic diversity

There are few reports of the tree being selected on the basis of fruit quality and cultivated commercially. However, Sundaresan (personal communication 1978) states that there are 3 'varieties' in Fiji but there is no description of these available. Various reports from Papua New Guinea suggest that there are quite a number of forms which can be recognized by the natives. Whether or not these forms are those referred to by Anon. (1977) is not known. Jacobs (1962) refers to 8 forms based on fruit colour — from dull yellow to purple. The best forms in Papua New Guinea are reputed to be in the Sepik district.

Production season and time of cropping

The season of fruit maturity in Fiji is from the end of November to early March. A similar season would be expected in northern Australia. In Florida, Whitman reports flowering in September and harvest in December. Southern Queensland would possibly have a winter harvest.

Sundaresan suggests trees commence bearing in 6 to 7 years from seed in Fiji whereas Whitman states 9 to 10 years in Florida.

Yield potential

The only data available are Sundaresan's comment that Fijian trees bear 700 to 1000 fruits per season.

Market acceptability

Whitman (1972) suggests the fruit in Florida tastes like bland longan and is pleasant but not outstanding. Fijian residents place moderate value on the crop. Kebby (personal communication 1977) rates certain forms in Papua New Guinea as good as lychee or rambutan. The north-west coast of Papua New Guinea is suggested as having the best tasting forms.

However, Jacobs (1962) states that in the Papua New Guinea-Solomons group, purple fruit forms are most highly regarded by the natives, and Paijmans *et al.* (1976) quotes it as the favourite fruit of the Sengseng people of New Britain. The Sepik River forms are also very good and are reputed to have a brittle shell, easily broken in the hand. Trees are commonly planted by Chinese residents of Rabaul and the "hardskin" types are preferred. These isolated trees do not appear to be excessively vigorous when compared with those growing naturally in the forests (J. Marshall personal communication).

Export potential

Export production of the fruit is unlikely.

Cultural details

Propagation

All trees in cultivation are seedlings but should clones be selected then grafting methods used for other sapindaceous species should be applicable.

Management

Due to the rapid growth of the tree, it is probably susceptible to damage by strong winds and should be planted in a sheltered situation. No further information is available.

Irrigation

The tree is native to wet tropical areas and where there is a prolonged dry season tree growth would obviously respond to irrigation. The effect of a dry period of flower initiation, fruit set and development is unknown.

Pests and diseases

In Florida, Whitman (1972) reports no injurious insects or diseases. In north Queensland the black beetle (*Rhyparida* spp.) attacks summer and autumn growth.

Harvesting methods

In Fiji, the Solomon Islands and Papua New Guinea ripe fruit are harvested by simply knocking the panicles with a long pole. This often involves tree climbing.

Transport characteristics

The fruit stands up well to transport due to its thick skin.

References

Anon. (1977), Taun — wood properties and uses. CSIRO Division of Build. Res., Inf. Service Sheet 10.

Paijmans, K., Van Balgooy, M.M.J. and Powell, J.M. (1976), New Guinea Vegetation. CSIRO in assoc. with Aust. Nat. Univ. Press., Canberra.

Jacobs, M. (1962), A study in variability of the genus *Pometia. Reinwardia* **6,** 6.

Whitman, W.F. (1972), The olosapo, the sunsapote and the Fijian longan. *Proceedings of the Florida State Horticultural Society* **84,** 323.

MAMONCILLO

by B.J. WATSON

Melicoccus bijugatus Jacq.
Melicocca bijuga L.

Botany

In the family Sapindaceae the genus *Melicoccus* contains only two species, both of which are native to the Caribbean region of tropical America. *Melicoccus bijugatus* Jacq. is considered to be the correct name for the species under consideration, although it is known in much of the literature as *Melicocca bijuga* L. Other common names are Spanish lime, quenepa, honeyberry, genip, anoncillo, Jamaica bullace plum, kanappy, knepe, knippleboom, maco and takeboom. It is 'cultivated' mainly in Cuba, Haiti, Jamaica, Puerto Rico, Grenada and the Florida Keys.

The mamoncillo is a slow growing, large, upright (10 to 18 m) grey trunked tree with ornamental value. Leaves are alternate-compound, with 4 leaflets, light green to blue-green.

As with Sapindaceae generally, flowers of the mamoncillo are borne on terminal panicles. The flowers are small and greenish white. It is now accepted that from a seedling population there arise two types: male or staminate trees, and bisexual (hermaphrodite) trees. Bisexual trees may produce some functionally staminate flowers, but the majority are functionally female — the stamens producing no pollen. Cross-pollination between trees is generally necessary for fruit set and isolated trees seldom produce fruit. Best fruiting occurs when a number of seedlings are grown together, there being a range of flowering time in staminate trees (providing pollen) corresponding to receptiveness of bisexual (functionally female) trees.

Fertilized flowers develop into dense clusters of small green, round or oblong fruit (20 to 40 mm in diameter and weighing from 9 to 30 g each). When mature, the fruit have a thin, tough, leathery pericarp surrounding a juicy, white, cream or light orange translucent pulp (the aril). The pulp flavour ranges from sweet to sour and juiciness is also variable. The single oval seed is edible after roasting. Fruits mature in 3 to 5 months from anthesis according to 'cultivar' characteristics and climate.

Climatic requirements

In Florida mature trees will only tolerate brief exposure to temperatures 1 to 2°C below freezing and young trees are killed with any fall below 0°C. The tree only grows satisfactorily on the Keys, the warmest winter area in the United States (24 to 25°N). In Cuba, Puerto Rico, Nicaragua, El Salvador and Venezuela, the main production is confined to lower elevations. The tree is apparently tolerant of a wide range of tropical climatic conditions but appears particularly adapted to drought in coastal sands and other light soils (Campbell 1976). Whitman (1977) states that rain during winter prior to the normal flowering period in spring in Florida results in a poor flowering. Thus a dry season in winter and spring would appear to be an advantage.

Soil requirements

The tree tolerates a very wide range of soil types from heavy clay loams to coastal sands to rocky (including calcareous rocks) soils. Tolerance to alkaline soils (pH 7 plus) is good.

Genetic diversity

Limited programmes of plant selection have been undertaken in Florida (Campbell 1976) and Puerto Rico (Jackson 1967). Main efforts have been to select material with the following characteristics: heavy and regular yields; large fruit, thick aril, small seed; aril with low to medium acidity, high sugar content, good flavour, moderate juiciness, low astringency, easy pulp separation from seed; thick pericarp with low incidence of splits. Campbell (1976) and others rate Newcomb, Montgomery, Cuban No. 1, Cuban No. 2 and Queen as having the best overall characteristics.

Production season and time of cropping

In Florida, earliest flowering occurs in April (spring) and harvest follows approximately 4 months later, giving a season from August to October (Campbell 1976). Puerto Rican trees mature fruit from July to September (Kennard and Winters 1960). The time of ripening in north Queensland could therefore be expected between January and April, depending on latitude and altitude of the growing area.

Market acceptability

The fruit is sold in the local markets of Cuba, Puerto Rico and other Caribbean and central American countries. So many seedling trees have poor eating characteristics that only the best selections appear to have potential for general acceptability.

Export potential

Export production is very unlikely in the short term.

Cultural details

Propagation

The difficulty of propagating selected material has restricted development of clonal selection. Veneer grafting and chip budding has proved consistently unsuccessful and approach grafting is the most commonly acceptable method (Campbell 1976). However, Whitman reports successful side veneer grafts where the scion is tied intermittently, allowing air circulation into the scion, but with the graft enveloped by a clear plastic skirt to keep water out. Malo (1970) suggests growing seedlings in a light-weight mixture and hanging up in the tree for approach grafting. Air layering of large limbs (greater than 50 mm in diameter at cincture point) is moderately successful.

Considering the current problems with detached scion grafting, it may well be worth attempting the modified Forkert bud graft which is commonly used with another sapindaceous species, the rambutan.

Management

The need to provide male functional flowering trees for pollination may possibly be avoided by treating some flowering panicles on female functional trees with 10 to 50 ppm (a.c.) of N.A.A. In rambutan this treatment induces development of male functional flowers with sufficient pollen for a successful set.

Irrigation

In low rainfall zones, irrigation during fruit development and the new lateral growth phase may be beneficial. However irrigation in autumn and winter is undesirable since (as with lychee) induced vegetative growth in this period precludes normal seasonal flowering.

Harvesting methods

Whole panicles are removed at maturity but no further information is available.

Marketing characteristics

Fruit are either detached singly from the rachis or presented as bunches intact. The pericarp of the better cultivars is thick enough to prevent superficial damage. The following table denotes characteristics of some Florida selections.

Characteristics of Florida honeyberry selections (Campbell 1976).

Selection	Fruit g	Pericarp g	Seed g	Pulp g	Pulp % of total
Arec 1	12.88	3.35	3.21	6.32	49.1
Queen	26.37	6.56	5.15	14.06	55.6
Newcomb	16.81	4.62	4.09	8.11	48.2
Montgomery	23.03	5.54	5.64	11.85	51.5

Transport characteristics

Mamoncillo stands up well to transport.

References

Campbell, C.W. (1976), Selection and propagation of the Spanish lime in Florida. *Proceedings of the Florida State Horticultural Society* **89**, 227.

Jackson, G.C. (1967), Promising selections of the honeyberry (*Melicocca bijuga* L.) from Puerto Rico. *Journal of Agriculture of the University of Puerto Rico* **51**, 66.

Kennard, W.C. and Winters, H.F. (1960), Some fruits and nuts for the tropics. U.S. Dept. Agric. Misc. Publ. 801, p. 86.

Malo, S.E. (1970), Propagation of the mamey sapote. Proceedings of the Tropical Region, American Society of Horticultural Science, **14**, 165.

Martin, F.W., Almeyda, N., Ruberte, R., and Jordan, F. (1976), Inventario de frutas tropicales en Puerto Rico. Fac. de Ciencias Agricolas, University of Puerto Rico.

Whitman, W.F.C. (1977), The Spanish lime. Rare Fruits Council Int. Inc. Data Sheet.

52 Lychee

53 Lychee

54 Lychee

55 Rambutan

56 Pulasan

57 Akee

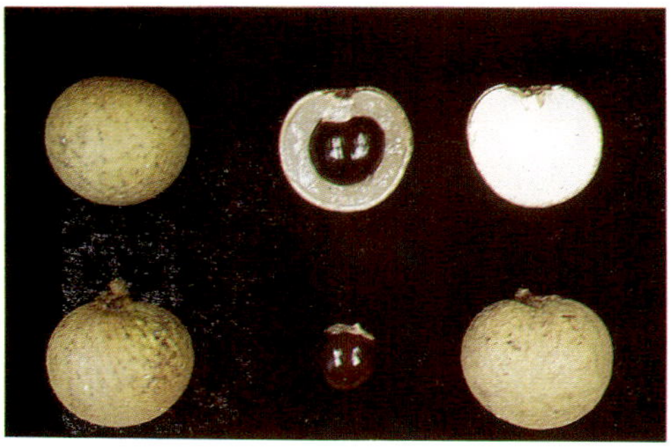

58 Longan

59 Rambutan

60 Rambutan

61 Rambutan

62 Sapodilla

63 Sapodilla

64 Mamey Sapote

65 Mamey Sapote

66 Caimito

67 Canistel

68 Grape — Lady Patricia

Chapter 27

SAPOTACEAE

SAPODILLA (CHIKU, CHICOSAPOTE, NISPERO, NASEBERRY, SAWO) by P.B. SCHOLEFIELD
Manilkara zapota (L.) Van Royen
Achras sapota L.; Manilkara zapotilla (Jacq.) Gilly; M. achras (Mill.) Fosberg.

Botany

The sapodilla is native to Central America and is an evergreen tree which can grow to 20 m high. Glossy leaves (5 to 13 cm long) are usually clustered at the end of small branches. Small, inconspicuous white flowers are borne in the leaf axils. The fruit is variable, round, oval or conical and 5 to 10 cm in diameter with a thin, brown, scurfy skin and yellow-brown translucent flesh which is soft, sweet and sometimes gritty. There are up to 12 black, shiny seeds per fruit.

Climatic requirements

The climatic distribution extends up to 30° north and south of the equator. It can tolerate the frosts often encountered in southern Florida but it generally grows best at medium to low elevations in tropical areas. The tree is reasonably hardy and is successfully grown in the drier areas of India and the Philippines on 1000 mm rainfall.

Soil requirements

Sapodilla prefers well-drained, light to medium textured soils of acid to neutral pH, but it can grow well on a wide range of soil types.

Production season

The main harvest period in Australia appears to be September to November with some fruit maturing several months either side of this. Because of the flowering and fruiting behaviour, small numbers of fruits can be found on trees at most times of the year.

Yield potential

Grafted trees start to produce significant crops after 5 years in the field. Yields reported from mature trees are:

Florida	—	20 to 30 tonnes per hectare
India	—	20 to 80 tonnes per hectare
Philippines	—	20 to 35 tonnes per hectare

No yields are available for Australia.

Market acceptability

The Australian consumer is not familiar with the sapodilla and its appearance (brown, scurfy skin) holds no instant appeal. However the fruit quality inside is excellent with the melting flesh having a rich, sweet flavour. The fruit is sold mainly as a fresh product although the ripe flesh can be pulped and used in desserts.

The season of maturity (September to November) is ideal for marketing in Australia because there is a shortage of soft fruits at this time.

Export potential

The fruit is well known in most Asian countries and good quality fruit should find ready acceptance on these markets.

Genetic diversity

Considerable diversity exists in seedling populations of sapodilla. Programmes of seedling assessment and selection have been carried out in Florida, India and the Philippines and most of the named varieties are from these countries. Some of these are: Prolific and Brown Sugar (Florida); Kalapatti, Cricket Ball and Dwarapudi (India); Pineras, Ponderosa and Sawo Manilla (Philippines). In Indonesia, Thailand, Malaysia and Singapore, some varieties exist but little information is available about them.

Some sapodilla cultivars

Variety	Origin	Fruit size	Fruit shape	Skin characteristics	Flavour
Prolific	Florida	small-medium	ovate	light brown	medium sweet
Brown Sugar	Florida	small	spherical-ovate	light brown	very sweet
Modello	Florida	medium-large	elliptical-ovate	light brown	medium sweet
Kalipatti	India	small	ovate-globose	light cinnamon	sweet
Cricket Ball	India	very large	round	—	very sweet
Dwarapudi	India	very large	round	—	very sweet
Pineras	Philippines	small	—	—	—
Ponderosa	Philippines	very large	elliptical-globose	—	—
Sawo Manila	Indonesia	medium	elliptical	light brown smooth	very sweet
Makok	Thailand	small-medium	round	light brown	very sweet
Krasuey	Thailand	small	elliptical	light cinnamon	very sweet
Kai Hanu	Thailand	medium-large	elliptical	—	sweet

Cultural details

Propagation

For commercial production, only vegetative propagation should be considered because of the slow growth and long period of fruiting of seedling sapodillas. Considerable research on types of rootstocks has been done in India and Sri Lanka. The related species *Manilkara hexandra, Manilkara kauki, Madhuca longifolia* and *Madhuca latifolia* have been used but because of soil differences, recommendations for Australia conditions cannot yet be made. Vegetative reproduction by grafting and budding is successful (Rowe-Dutton, 1976).

Management

Plantings are usually made at approximately 7 m x 7 m with wider spacings on more fertile soils. Little information is available on pollination requirements.

Irrigation

Young trees require regular irrigation and mature trees will benefit from supplementary irrigation during the dry season.

Fertilization

The sapodilla is not considered to require heavy fertilizing and moderate quantities of NPK mixtures, perhaps in split applications, should suffice.

Pests and diseases

The pests recorded in Florida include the larvae of *Eucosmophora* sp., which is a 'leaf miner' pest, various scale insects and mealy bugs. The caribbean fruit fly attacks the fruit making it unusable.

The fungal diseases, leaf rust *(Uredo sapotae)* and leaf spot (*Septoria* sp.), can cause serious leaf damage.

In India and the Philippines, galls are reported at the junctions of branches which retard growth and may kill the branches. The cause is unknown but it is described as being 'physiological'.

Harvesting methods

Assessment of fruit maturity is difficult because there is no marked colour change of the brown skin. The fruits are picked before they soften, either by hand or with picking poles. Some latex exudes from the pedicel and the fruit may require washing before packaging.

Marketing characteristics

None of the markets in Australia is familiar with the fruit and there would need to be a programme of controlled introduction and education.

Storage characteristics

In most of the areas of production the fruit is sold nearby and little storage is needed, but it has been reported that sapodillas can be stored either soft-ripe or hard (see Broughton and Wong 1979).

Transport characteristics

When picked hard-mature, the fruit takes 5 to 7 days to soften at room temperature. Therefore they are similar to the avocado and can be transported easily to market and soften after being sold.

Processing characteristics

The fruit are mainly eaten fresh but the ripe pulp can be used in sherbets and conserves. The flesh is low in acids.

References

Broughton, W.J. and Wong, H.C. (1979), Storage conditions and ripening of chiku fruits *Achras sapota* L. *Scienta Hort.* **10,** 377.

Rowe-Dutton, P. (1976), *Manilkara achras* — Sapodilla, *in* R.J. Garner, and S.A. Chaudri (eds), The propagation of tropical fruit trees. *CAB Hort. Review* **4,** 475.

Further reading

Coronel, R.E. (1976), Growing of chico. Univ. Philip. Los Banos. Ext. Circ. 12.

Randhawa, G.S. and Kohli, R.R. (1966), Sapota cultivation in India. *Indian Hort.* **10,** 3.

Ruehle, G.D. (1951), The sapodilla in Florida. Univ. Florida Ag. Exp. Sta. Circ. S-34.

Watson, B.J. (1978), An appreciation of five exotic tropical fruit tree species and their potential in Queensland. Queensland Department Primary Industries, Kamerunga.

STAR APPLE (CAIMITO) by P.B. SCHOLEFIELD
Chrysophyllum cainito L.

Botany

The star apple is native to the West Indies and Central American regions. It is a very attractive tree to 15 m high with dark green leaves (8 to 15 cm x 5 cm) which have a silky, golden-brown undersurface. The small purplish white flowers are borne in axillary clusters on mature twigs. Fruits are green, purple or copper coloured, round, up to 10 cm in diameter with a smooth skin containing a white, sweet pulp arranged in eight segments surrounding the hard, brown, glossy seeds. When halved, these segments give the fruit a star-shaped pattern.

Climatic requirements

The tree is tropical in its requirement and less tolerant of low temperatures than the sapodilla.

Soil requirements

The star apple is reported to grow well in a range of soils from sands to clay-loams and should not be limited in Australia by soil type.

Genetic diversity

Great diversity exists in the many seedling types grown. Essentially there are two races or types with a purple or pale green skin of the fruit. The purple fruit is reported to have more flavour while the green fruit is sweeter. Named varieties include Publico, Haitian, Weeping and Florida.

Production season and time of cropping

The reported times of cropping vary from early spring, late spring to midsummer. Two crops per year have been reported for some varieties.

Yield potential

The only report of yield (Ingram 1976) is of Indian trees producing 70 kg per tree.

Market acceptability

The fruit should find ready market acceptance although it is considered to be low in acid, and if not completely ripe, contains some latex.

Export potential

The fruit is well known in many Asian countries.

Cultural details

Propagation
Considerable work in the Philippines has found the tongue-inarching method most successful, while others report success with marcotting and cutting propagation.

Harvesting methods

The fruits are allowed to fully ripen on the tree and are picked by hand.

Further reading
Ingram, M.H. (1976), *Chrysophyllum cainito* — Star apple, *in* R.J. Garner, and S.A. Chaudri (eds), The propagation of tropical fruit trees. *CAB Hort. Review* **4**, 314.
Popenoe, W. (1920), 'Manual of tropical and sub-tropical fruits'. Macmillan, New York, p. 344.

MAMEY SAPOTE
Pouteria sapota (Jacq.) H.E. Moore and Stearn

by P.B. SCHOLEFIELD

Botany

Like most of the other fruits of this family the mamey sapote originated in the Central Americas. The tree is large (up to 30 m high) with shiny, dark green leaves 14 to 30 cm long arranged compactly on the shoots. Small sessile white flowers are borne on the bare branches below the tufts of leaves. The fruit is ellipsoidal or ovoidal, 10 to 25 cm x 8 to 12 cm and may weigh up to 3 kg. The brown skin is thin, tough and rough, enclosing pink to red flesh which is soft, sweet and aromatic when ripe. One to three seeds (5 to 6 cm long) are usual.

Climatic requirements

The tree is strictly tropical and cannot withstand low temperatures. It is being grown in Florida but can be expected to survive only in the warmest areas.

Soil requirements

Many soil types are suitable from the heavy clays of Puerto Rico to the limestone soils of Florida. Good drainage is essential as the trees cannot withstand waterlogging.

Genetic diversity

The mamey sapote has been propagated mainly by seed with a great diversity in tree and fruit characteristics. Several named varieties have been selected: 'Magana', 'Cuban No. 1' and 'Cuban No. 2'. Problems with propagation have hampered the development of superior selections.

Production season and time of cropping

Fruits mature during the summer months but the trees can produce fruits over long periods.

Market acceptability

The fruit is greatly prized by the people of Central America and the impetus for commercial expansion in Florida has been the influx of Cuban immigrants. Some market education would be necessary in Australia.

Cultural details

Propagation

Propagation problems have limited the culture of the fruit. The side veneer graft is the method most commonly used. Grafted trees bring very high prices in commercial nurseries in Florida because of the big demand for trees.

Management

The trees grow very large and planting distances need to be at least 8 m x 8 m, possibly up to 12 m x 12 m.

Irrigation

Mamey sapote is highly susceptible to short periods of drought and will drop leaves if stressed. Irrigation is therefore essential in the dry season.

Fertilization

Fertilizer application similar to those for mango and citrus have proved satisfactory.

Pests and diseases

Pests and diseases are few but the sugarcane root borer *(Diaprepes abbriviatus)*, scale insects, spider mites and termites have caused problems.

The fungal diseases anthracnose *(Phyllosticta sapotae)* and rust *(Uredo sapotae)* can cause damage to leaves.

Harvesting methods

Harvesting can be difficult because of the large size of some trees. Hand harvest is used and the fruit is picked mature but firm when it begins to redden.

Marketing characteristics

No established marketing procedures exist for most of the Sapotaceae fruit in Australia.

Processing characteristics

In the Caribbean, drinks, jellies and pastes are made from ripe fruit and a bitter chocolate can be prepared from the seeds.

Further reading

Almeyda, N. and Martin, F.W. (1976), Cultivation of neglected tropical fruits with promise. 2. The Mamey sapote. USDA, ARS-S-156.

Campbell, C.W. (1967), The mamey sapote in southern Florida. *Proceedings of the Florida State Horticultural Society* **80,** 318.

Popenoe, W. (1920), 'Manual of tropical and sub-tropical fruits'. Macmillan, New York, p. 340.

GREEN SAPOTE (INJERTO)

Calocarpum viride Pittier

by P.B. SCHOLEFIELD

Botany

The green sapote is indigenous to the Guatemalan highlands and Honduras. The tree resembles the mamey sapote but has a smaller leaf, pubescent on the undersurface. The fruits are up to 12 cm long with a brownish green to yellow-green, thin skin. The red-brown flesh is sweet, melting and juicy and contains one or two large seeds.

Climatic requirements

They are thought to be slightly more cold tolerant than the mamey sapote, but similar in other requirements.

Soil requirements

Soil requirements for green sapote are the same as for mamey sapote.

Genetic diversity

Little information is available and there are no commercial varieties.

Production season and time of cropping

In Florida the fruiting season is from December to March. Seedling trees can take long periods to come into bearing.

Market acceptability

The green sapote is reputedly an excellent quality fruit.

Cultural details

Propagation

Grafted trees should be used, by side grafting onto itself or mamey sapote seedlings.

Harvesting methods

The green sapote is hand harvested when firm-ripe.

Further reading

Popenoe, W. (1920), 'Manual of tropical and sub-tropical fruits'. Macmillan, New York, p. 343.

Whitman, W.F. (1965), The green sapote, a new fruit for south Florida. *Proceedings of the Florida State Horticultural Society* **78**, 330.

CANISTEL
Pouteria campechiana (HBK) Beahni

by P.B. SCHOLEFIELD

LUCMO
Pouteria obovata HBK

ABIU
Pouteria caimito (Ruiz and Pav.) Radlk.

Canistel, lucmo and abiu are closely related and will be considered together. Martin and Malo (1978) give an extensive list of botanical synonyms and common names.

Botany

CANISTEL
The species is native to the middle Americas and trees may reach 20 m in height. The dark green, shiny leaves range from 12 to 25 cm in length and flowers are small, pale green to white. The fruit is ovoid, pointed at the stigma end, with a smooth, yellow to orange skin enclosing a mealy, yellow pulp with several large, shiny seeds. Fruit size is 4 to 5 cm in diameter, 5 to 17 cm long with a weight of up to 1.5 kg.

LUCMO
Lucmo is similar to canistel but generally with larger leaves and fruit. It is native to Ecuador, Peru and northern Chile.

ABIU
A native of the Amazon headwaters area, this tree is slightly smaller than the above species. The fruit is very different, 8 to 15 cm long with a bright yellow skin and sweet white translucent flesh containing 2 to 3 oblong seeds.

Climatic requirements

This group of species are widely adaptable and can grow in most parts of the tropics and subtropics, provided the areas are frost free. The canistel and lucmo can withstand drier climates with lower rainfall. The abiu grows better in warm humid areas, similar to its area of origin.

Soil requirements

The trees can grow on a wide variety of soils, provided drainage is good.

Genetic diversity

Great seedling diversity exists and some selections have been made but few named varieties exist.

Production season and time of cropping

Seedling trees may take 3 years (canistel) to 8 years or more (lucmo) to start bearing fruit. The fruits are reported to mature in December to March in Florida with little information from other areas.

Market acceptability

Market acceptability is yet to be determined but the abiu may be more acceptable than the canistel or lucmo.

Cultural details

Propagation

Grafting is recommended and the types of graft commonly used are side veneer, chip bud and cleft graft. The abiu is more difficult to propagate than the canistel or lucmo.

Harvesting methods

Fruit are harvested by hand when firm but mature. Experience is needed to determine the correct stage for picking.

Marketing characteristics

None of the fruit has been marketed in commercial quantities anywhere in the world.

Storage characteristics

Fruits do not store well.

Processing characteristics

The canistel and lucmo have high nutritive value. The flesh of these two fruits can be processed into a dried meal which can be added to desserts and ice cream.

References

Martin, F.W. and Malo, S.E. (1978), Cultivation of neglected tropical fruits with promise. 5. The canistel and its relatives, USDA S.E.A. New Orleans.

Further reading

Popenoe, W. (1920), 'Manual of tropical and sub-tropical fruits'. Macmillan, New York, p. 347.

MISCELLANEOUS SPECIES

by P.B. SCHOLEFIELD

There may be other members of the Sapotaceae family that warrant inclusion in this list. One of these is the miracle fruit, *Synsepalum dulcificum,* which when eaten, affects the taste buds making it possible to eat very acid fruits without any discomfort.

Chapter 28

VITACEAE

GRAPES

by P.B. SCHOLEFIELD

Vitis sp.

The genus *Vitis* is here divided into five groups:

1. European grapes, *Vitis vinifera.*
2. Muscadine grapes, *Vitis rotundifolia.*
3. Labrusca grapes, *Vitis labrusca* hybrids.
4. 'French hybrids' — this group of hybrids originated from hybridization of a wide range of American species with *V. vinifera.*
5. *Vitis caribaea* hybrids.

Botany

V. vinifera (L) is believed to have originated in Europe to western Asia. Great variability exists in this species but it generally has large, glabrous, lobed leaves. Bunches are large and berries of medium to large size have a thin skin and flesh without the musky flavours associated with the American species.

V. rotundifolia Michx: The muscadine grape has its origin in south-east North America. The leaves have prominent serrated edges and glossy upper and lower surfaces. Bunches are small (3 to 20 berries) with large individual berries 12 to 30 mm in diameter. The berries absciss from the pedicel when ripe, have a thick tough skin, are seeded, and have flesh with a musky flavour.

V. labrusca (L) hybrids: The labrusca type grapes were bred in the north-east United States. Vines grow vigorously with shoots generally having a tendril at each node and large downy leaves. Small bunches have large berries with a thick skin and a musky flavoured flesh.

'French hybrids': These grapes are hybrids between *V. vinifera* and a range of American grape species. They were produced by breeders in France in an attempt to incorporate the resistance to phylloxera with the grape quality of *V. vinifera,* following the devastation of European vineyards by phylloxera last century.

V. caribaea (L) hybrids: A vine originating in the warm Caribbean area. The leaf has a woolly undersurface. The bunches are small with small berries. Only a few hybrids exist but because of its area of origin, these should be considered for tropical areas.

Climatic requirements

Grapes are grown over a wide climatic range, from areas with very severe winters in eastern Europe and Canada, to the lowland tropical areas of Thailand and Venezuela. Different varieties are suited to the various climatic regions. However, as a general rule, grapes thrive best under hot, dry conditions during the period from flowering to fruit maturity, provided sufficient irrigation water is applied.

In the tropics, the normally deciduous grapevine is evergreen with no dormant period and grows continually. Some vines exhibit considerable bud dormancy giving poor budburst. Pruning management can partially overcome this problem.

The seasonal distribution of rainfall is more important than the total amount. Rain at fruit maturity causes splitting of berries. Conditions of high humidity are conducive to the development of fungal diseases of leaves and fruit which can drastically reduce yields. The American species are generally more able to tolerate conditions of high humidity.

Soil requirements

Grapes are adapted to a wide range of soils from sand to clay-loam, but those that are light and well drained are most suitable. Soil pH levels of 5 to 8 are satisfactory with pH 6.5 to 7.5 being preferred.

Genetic diversity

Great diversity exists in all the five groups of grapes. Fruit characteristics range from very seedy to seedless; black, red or white skin colour; large or small, spherical or elongated berries. Growth habits and resistance to diseases vary greatly.

Some varieties reported as growing in tropical areas:

- Vinifera: (most of the Indian varieties are vinifera types). Anab-e-shahi, Cheema Sahebi, Gulabi, Kali Sahebi, Perlette, Black Champa, Pandhari Sahebi, Muscat Hamburg, Cardinal, Ribier.
- Muscadine: Southland, Magoon, Chief, Dixie, Carlos, Welder (all Florida varieties). Others bred in Georgia and Carolina.
- Labrusca: Lake Emerald, Blue Lake, Stover (Florida varieties). Isabella, Iona, Wilder, Bangalore Blue, Bangalore Purple.
- French hybrids: Villard Blanc, Villard Noir.
- Caribaea: Criolla Negra and other criolla types of Venezuela. Some hybrids bred in Costa Rica.

Some tropical grape cultivars

	Berry size	Berry colour	Seeds	Maturity	Yield
(a) **Vinifera**					
Anab-e-Shahi	large	white	seeded	late	heavy
Perlette	medium	white	seedless	early	medium
Cardinal	large	red	seeded	early	medium
Ribier	large	black	seeded	late	heavy
(b) **Muscadine**					
Southland	very large	black	seeded	—	medium
Magoon	very large	black	seeded	—	medium
(c) **Labrusca**					
Isabella	medium	purple	seeded	—	—
Bangalore Blue	small-medium	purple	seeded	—	medium
(d) **French hybrids**					
Villard Blanc	medium-large	white	seeded	mid-season	heavy
Villard Noir	medium	black	seeded	early-mid	medium
(e) **Caribaea**					
Criolla Negra	II	black	seeded	continuous*	heavy

* Bears successive crops on laterals of increasing order.

Production season and time of cropping

In the tropics the production season is not determined as much by climate but rather by management, particularly the time of pruning. However, management systems will have to be developed for the various areas to control the period of fruit maturity. Ideally this should coincide with the shortage of supply to markets in the spring months, and avoid the 'wet' season.

Yield potential

Grapes can produce crops in 2 to 3 years from planting, a shorter period than many other perennial horticultural crops.

Some reported yields from tropical areas are:

Vinifera — Anab-e-Shahi in India, 29 to 37 tonnes per hectare
Muscadine — Florida varieties, 7 to 17 tonnes per hectare
Labrusca — Florida varieties, 7 to 15 tonnes per hectare
French hybrids — Villard Blanc, Villard Noir, approximately 20 to 30 tonnes per hectare
Caribaea — up to 40 tonnes per hectare

However, quality is important with grapes for table consumption and often bunch-thinning may reduce the yields, but increse the returns.

Market acceptability

In the tropical areas of Australia I consider that the development of a wine industry is unlikely and that the main market will be for table grapes.

The acceptability of grapes from tropical production areas will depend greatly on the season of production. The vinifera grapes are more acceptable than the musky flavoured American species although some of the French hybrids have little of this flavour.

Export potential

The export potential for grapes from the tropical areas of Australia to Asia appears good, but market investigations will need to determine the preferences of these markets.

Cultural details

Propagation

Propagation is either by cuttings or grafting onto suitable rootstocks.

Management

Most of the production areas in the tropics are in countries with adequate cheap labour. Therefore their management systems will not be directly applicable to Australian conditions.

Trellising of the vines will be necessary. Some muscadine and labrusca varieties have female flowers and require a nearby pollen source. These varieties should be avoided. Some vinifera table grapes (for example, Ohanez, Nimrang) also have female flowers.

Irrigation

Irrigation is essential in the dry season and adequate water is very important at the critical stages of flowering and fruit set.

Fertilization

Addition of lime may be necessary to increase the pH on some acid tropical soils. Although grapevines do not have a high fertilizer requirement the summer rainfall may leach nutrients. Mixtures high in K are recommended. Excess N is to be avoided because it encourages vigorous shoot growth to the detriment of cropping.

Pests and diseases
The main fungal diseases are downy mildew *(Plasmopara viticola),* black spot or anthracnose *(Gloeosporium ampelophagnum),* and powdery mildew *(Uncinula necator).* Insect pests can be many including leaf-eating caterpillars, beetles, and mealy bug. Several virus diseases (leaf-roll and fan-leaf) occur but fan-leaf is not a problem in Australian vineyards. Pierce's disease devastates vines in south-east United States. Plant parasitic nematodes can be a serious problem in light soils.

Harvesting methods

Hand harvesting is necessary to produce quality table grapes but several harvest aids can reduce the costs of picking and packing. Mechanical harvesters are commercially available but are mainly used for wine grapes.

Marketing characteristics

All markets in Australia are familiar with the handling of grapes. However varieties that are greatly different from those normally grown would encounter some market resistance.

Storage characteristics

The technology for cool storage of grapes for long periods (up to 6 months) is available and being used in Australia. However whether the varieties described here will be suitable for long-term storage is not known.

Transport characteristics

The potential tropical grape growing areas are a long distance from markets. Any new varieties will have to be assessed for their transportability. Thick skins, firm flesh and pedicels that do not dry out are necessary attributes. Pre-cooling before transport is essential.

Processing characteristics

Although grapes can be sold fresh, dried, or for wine, it is envisaged that fresh grapes for table use will be the main outlet.

Further reading
Anon. (1977), Grape cultivation. Ext. Bulletin 15, I.C.A.R. Indian Inst. Hortic. Res., Bangalore.
Aubert, B. (1972), Viticulture in tropical areas for table grape production: aspects and possibilities. *Fruits* **27,** 513 (in French).
Crocker, T.E. and Mortensen, J.A. (1976), The muscadine grape. Florida Coop. Ext. Service Bulletin FC-16.
Crocker, T.E. and Mortensen, J.A. (1976), The bunch grape. Florida Coop. Ext. Service Bulletin FC-17.
Vega, J. (1971), Grape culture in tropical countries. *Office Int. Vigne Vin. Bull.* **44,** 775.

INDEX